River Basin Management - Sustainability Issues and Planning Strategies

Edited by José Simão Antunes Do Carmo

Published in London, United Kingdom

IntechOpen

Supporting open minds since 2005

River Basin Management - Sustainability Issues and Planning Strategies
http://dx.doi.org/10.5772/intechopen.91591
Edited by José Simão Antunes Do Carmo

Contributors
Patrick Ombiono, Ozekeke Ogbeide, Azubuike Chukwuka, Rocky Talchabhadel, Kenji Kawaike, Hajime Nakagawa, Innocent Rangeti, Bloodless Rimuka Dzwairo, Luis Altarejos-García, Mario Andrés Urrea-Mallebrera, Juan T. García, Mak Sithirith, Marina García, Damelis Jáuregui, José Simão Antunes Do Carmo

First published in London, United Kingdom, 2021 by IntechOpen
IntechOpen is the global imprint of INTECHOPEN LIMITED, registered in England and Wales, registration number: 11086078, 5 Princes Gate Court, London, SW7 2QJ, United Kingdom
Printed in Croatia

British Library Cataloguing-in-Publication Data
A catalogue record for this book is available from the British Library

Additional hard and PDF copies can be obtained from orders@intechopen.com

River Basin Management - Sustainability Issues and Planning Strategies
Edited by José Simão Antunes Do Carmo
p. cm.
Print ISBN 978-1-83968-130-1
Online ISBN 978-1-83968-131-8
eBook (PDF) ISBN 978-1-83968-132-5

For EU product safety concerns:
IN TECH d.o.o., Prolaz Marije Krucifikse Kozulić 3, 51000 Rijeka, Croatia, info@intechopen.com or visit our website at intechopen.com.

Meet the editor

José Simão Antunes do Carmo obtained his master's degree in Hydraulics and Water Resources in 1990 from the University of Lisbon and his Ph.D. in Engineering Sciences in 1995 from the University of Coimbra, Portugal. He was director of several undergraduate and master's courses in Civil Engineering and Environmental Engineering in the period 1995-2010. He was a scientific advisor of twenty-five master's dissertations and two Ph.D. theses in Hydraulics and Water Resources. He has published two books, sixty-five papers in scientific journals, fifteen book chapters, and more than one hundred papers in international conferences. He has edited eleven books and a special issue of a scientific journal. He is a member of the editorial board of several international journals. His main areas of scientific research are hydrodynamics, morphodynamics, coastal management, river and coastal processes, climate change, natural hazards, risks, and vulnerabilities.

Contents

Preface

The dynamics of the inter-relationship between the various elements that constitute a hydrographic basin involve cause-effect relationships that can lead to constant alterations in fluvial dynamics. The combination of elements such as morphology, altimetry, and structural controls in a fluvial network enables a specific longitudinal profile to develop that is dynamic and constantly in search of an equilibrium between the rates of water discharge, erosion, transport, and deposition of sediment.

However, human activities carried out on a stretch of the river may alter this balance in different ways and with different levels of intensity. Indeed, worldwide, the morphology of fluvial systems has been dramatically altered by man's actions. Recent changes in water systems are the result of uses and practices that have been carried out without due care to provide suitable management of the natural resources.

It is widely recognized that the economic, social, and environmental conditions in several hydrographic basins and estuarine systems have changed dramatically in the last decades in consequence of anthropogenic activities, and they will go on changing in the forthcoming years due to increasing human pressure. It is also clear that human activity in these natural systems contributes to global warming and climate change, and it is certain that such consequences will be felt most strongly after the 2050s. Adjustments in natural and human systems are thus necessary to respond to the present-day effects and expected impacts of climate change.

The soils deterioration, changes in landscape structure and functionality, extraction of inert materials, building of hydraulic infrastructures, untreated urban and industrial effluent discharges, and unregulated hunting and fishing activities are all associated with a reduction in the quality of life of riverside populations, forcing us to think about the future in a different way.

Seeking to reverse such a course of degradation, arguments are put forward for development based on the relationship between people and nature, linking activities orientated towards economic dynamism with improvements in the quality of life of local populations, participation in political decision-making, preservation of the natural environment, and concern for the future.

New strategies are needed to develop a multifunctional use structure, which must take into account the multiple aims of sustainable development. Efficient management of water resources necessarily includes measures to strengthen institutions, support for citizen participation, and the development and transfer of technology for this purpose.

To support this development, integrated analyses of new sources of information are needed. In addition, it must be possible to carry out planning scenarios during the decision processes and, for this, we need instruments that assist decision-makers in the search for optimized solutions to existing problems.

Over the past thirty years, Geographical Information Systems (GIS) have developed considerably for Hydrology and Water Resources, and their use now extends far beyond land use planning and thematic geo-referencing for spatial analysis. Indeed, water resource information systems bring together information on quality, quantity, availability, and demand for water, as well as the factors – natural or resulting from human actions – that affect its use.

The data collected must be duly substantiated and remain accessible through a reliable and solid database from a scientific and technical point of view. The systems must also be dynamic in such a way that, on the one hand, they incorporate the information generated by the integrated management system and, on the other hand, they provide the necessary information for the definition of projects, activities, and interventions in river basins.

It must also be recognized that social participation is the most effective way to select the best path to follow in terms of conserving natural resources and guaranteeing their rational and democratic use. For this to happen, it is necessary that social actors and public agents are aware of the role that each must play.

Guaranteeing and exercising social participation are difficult tasks that demand space and adequate mechanisms to ensure that: (1) water management is oriented towards collectively established guidelines and goals, and that (2) projects based on user and community needs and priorities are economically efficient, socially effective, and do not harm the environment.

It is in this overall context that this book emerges, which aims to incorporate different disciplines and approaches, including theoretical analyses, field data, modeling tools, optimization techniques, and economic–ecological instruments in the management of watersheds and estuaries.

Contributions in this book cover the following topics:

- Fluvial processes and adaptation measures
- River and watershed management
- Flood and sediment management
- River dynamics and restoration
- Vulnerabilities and natural risks
- Water quality management
- Ecological perspectives
- Case studies

I believe this book comprises material of enough quality and quantity to make it a useful reference in the field of river systems.

José Simão Antunes Do Carmo
Department of Civil Engineering,
University of Coimbra,
Coimbra, Portugal

Acknowledgements

The editor is grateful to all authors for their excellent contributions, as well as to the invited reviewers for their assistance in reviewing all proposals and final chapters, thus ensuring the quality of the book. Special thanks are due to Ms. Dolores Kuzelj for being a diligent Author Service Manager, always available to help, and offer ongoing support. It has been a pleasure for me to coordinate this project.

Section 1

Institutional Vulnerabilities and Planning and Governance Strategies

Chapter 1

The Hintze Ribeiro Bridge Collapse and the Lessons Learned

José Simão Antunes Do Carmo

Abstract

In March 2001, a serious accident occurred in Portugal during a flood on the Douro River, next to Porto, Portugal. The collapse of the Hintze Ribeiro Bridge killed 59 people traveling in a bus and in three cars that fell into the Douro River. This bridge was built at the end of the 19th century on a curve of the Douro River, next to the mouth of the Tâmega River, approximately 50 km upstream of Porto. It was found that the combined effects of sand dredging in the 25 years prior to the accident (1975–2000) and the erosion produced by five consecutive floods between December 2000 and March 2001 were the main causes of this accident. Aiming to contribute to the prevention of occurrences such as that registered in Portugal with the collapse of the Hintze Ribeiro Bridge, a brief overview of this bridge is presented herein, as well as the causes that led to the collapse, some reflections on the processes involved, and, mainly, the lessons learned.

Keywords: Douro River, bridge collapse, floods, sediment transport, computational model, numerical simulations

1. Introduction

In March 2001, the collapse of the Hintze Ribeiro Bridge killed 59 people that were crossing the bridge at the Douro River [1] in a bus and three cars. The bridge collapse occurred on the night of 4 March 2001 when the Douro River flooded. The accident happened after the fifth river flood in a series of flood waves that started on 6th December 2000, all with peaks greater than 8000 m^3/s.

This bridge was built at the end of the 19th century over a curve of the Douro River, near the mouth of the Tâmega River. In the 1980s, a dam was built downstream (Crestuma–Lever), whose reservoir reached the bridge and began affecting its pillars. Upstream of the section where the Hintze Ribeiro Bridge was located are Carrapatelo Dam on the Douro River and Torrão Dam on the Tâmega River (**Figure 1**).

2. Background

The Hintze Ribeiro Bridge, also known as the Entre-os-Rios Bridge, was built over the Douro River between 1884 and 1886 to link the village of Entre-os-Rios to Castelo de Paiva. The bridge structure consisted of six pillars, with only two permanently exposed to the flow (pillars P2 and P3), while P4 and P5 were implanted

Figure 1.
An actual perspective of the entre-os-Rios village. The second bridge seen in this perspective did not exist in 2001. The Hintze Ribeiro bridge was located in the same section as the first bridge (the nearest) seen in this image (adapted from [2]).

Figure 2.
Photo of the Hintze Ribeiro bridge, dated 1931 [3, 4].

Figure 3.
*Photo of the Hintze Ribeiro bridge taken in the late 1970s before the construction of Crestuma–lever dam [5]. As in **Figure** 2, designations P1 to P6 were used for the pillars of the bridge starting from the right bank (in the figure, from left to right).*

on a sand deposit that existed until 1975, as shown in **Figure 2** [3, 4]. In addition, looking at **Figure 2**, it can be seen that P2 and P3 were protected by rockfill. The P3 protection is seen more clearly in **Figure 3**.

The analysis carried out on the various river bathymetric surveys that took place from the beginning of the 20th century until the 1970s allowed us to conclude that there were no bathymetric depressions along the stretch of the Douro River upstream of the Hintze Ribeiro Bridge. With the increase in sand extraction from the 1970s onward, the sand deposit around P4 and P5 pillars (**Figure 2**) began decreasing due to direct extraction, having almost disappeared by 1982, as reported in a survey carried out in that year. This is also clearly shown in **Figure 3**.

Between the 1970s and the construction of the Crestuma–Lever reservoir in 1980s, large amounts of sand were extracted directly from the sand deposit at the left river bank where the bridge was located, and, after filling the reservoir, sand extraction continued across the entire reservoir by means of boats and by dredging the riverbed of the Douro River along the stretch upstream of the bridge. In **Figure 3**, a photo of the Hintze Ribeiro Bridge taken in the late 1970s before the construction of Crestuma–Lever Dam, one can observe the sand deposit irregularity, the almost inexistence of sand around P4, and the absence of rockfill protecting this pillar.

The sand extractions carried out upstream of the bridge, up to 7 km, created large riverbed depressions that retained part of the transported sands, which were already visible in the river bathymetry (outline) shown in **Figure 4**. Thus, the sandbar that naturally formed on the Douro River curve where the Hintze Ribeiro Bridge was located was never rebuilt.

By analyzing the technical documentation regarding the surveys of the bridge along the time (unpublished reports), it is possible to observe a lowering of the river bottom, next to P4, of approximately 11.5 m in the period from 1913 to 1982, and of 1.5 m between 1982 and 1989 [3]. Considering that the inner margin of the Douro River curve where P4 was implanted is a natural deposition area (see **Figures 1** and **2**), the lowering of the bottom level does not reflect this natural tendency.

The flows occurring in the section of the Douro River under the Hintze Ribeiro Bridge can be quantified by adding the flows released by the Carrapatelo and Torrão dams and the flows from the intermediate basins, resulting in the flow values discharged into the Cestuma–Lever reservoir, as shown in **Figure 5**. This figure shows the hourly average flows released from the Carrapatelo and Torrão dams, as well as the instantaneous flows measured at the entrance to the Crestuma–Lever reservoir.

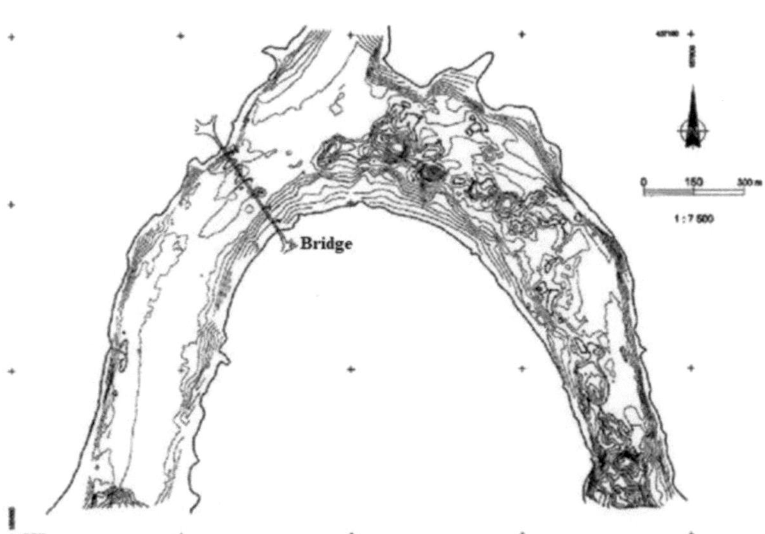

Figure 4.
Douro River survey in entre-os-Rios carried out in 1982 [3]. The depressions are outlined upstream of the bridge, mostly along the thalweg.

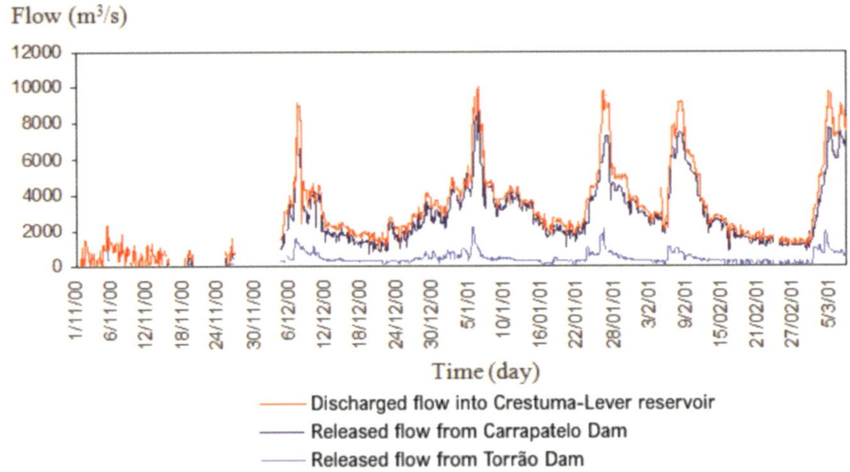

Figure 5.
Released flows from the Carrapatelo and Torrão dams and the flow discharged into the Crestuma–lever reservoir between 1 November 2000 and 7 march 2001 [3, 4].

This sequence of five intense floods, with river flows that remained above 2000 m^3/s for over three months, did not allow sufficient sediment deposition to guarantee the replacement of the previous bed conditions (former topography). In fact, in addition to generalized erosion of the river bed, these hydrodynamic conditions favor erosion around the foundations of structures (scour), as occurs for higher flows—in this case, approximately 2500 m^3/s, estimated from the knowledge of the particle size of the bottom material and the critical velocity for the beginning of sediment transport by entrainment. The duration of the average daily flows of these floods are shown in **Table 1** [6].

Thus, the riverbed lowering of the order of 4 m, on average, observed between 2000 and 2001 in the stretch of the Douro River where the Hintze Ribeiro Bridge was located was mainly due to the five major floods observed from November 2000 to March 2001. As the amount of sand transported was not enough to meet the needs generated by the five floods, due to the retention of sand in the depressions that were created upstream by dredging or extracting sand over more than two decades, it was not possible to restore the conditions prior to the floods.

A video made by a specialized company in 1986 and, subsequently, a geological survey carried out in September 1988 allowed the detection of some corrosion in the metallic protections of P2 and P3, as well as the lack of riprap protecting P4 (**Figure 2**). In addition, even in 1986, a marked decrease in the level of the river bottom was also detected, especially near P4. It was discovered that the sand was at level − 2.40 m (approximately 38 m below the bridge deck).

Floods	Number of consecutive days with average daily flows released from Carrapatelo Dam equal or greater than 3000 m^3/s
1 December 2000 until 4 March 2001	2 days (7 and 8 December 2000) 2 days (10 and 11 December 2000) 14 days (2–15 January 2001) 8 days (25 January–1 February 2001) 6 days (7–12 February 2001) 2 days (3 and 4 March 2001)

Table 1.
Average daily flow duration from 1 December 2000 to 4 march 2001 [6].

The survey carried out in 1988 showed that, in the case of the foundation of P4, the top of the metal campanula used for the excavation was located 43–44 m below the bridge deck, and the level of the river bottom was approximately 45.8 m below. This topo-hydrographic survey also made it possible to verify that the pillars were not supported on rock, but on alluvial soil.

In addition, analyzing the 1988 survey, it was also possible to conclude that at that time, there were approximately 9 m of sand above the foundation of P4. A new topo-hydrographic survey carried out 10 years later, in 1998, showed that the situation had worsened, as the bottom of the river lowered by approximately 1 m in the period 1988–1998.

As evidenced by the photo shown in **Figure 6**, taken a few days after the collapse of the bridge, P4 fell against the current and, therefore, against the effect of the natural flow, whose strength and resulting momentum would have pointed in exactly the opposite direction. Such behavior can only be explained by infra-excavation (scour), which was the dominant process in the foundation of P4, putting it out of service. In fact, the location and magnitude of the scour hole usually occurs in front and around of the structure [3, 4].

The Hintze Ribeiro Bridge, although old, had a construction design and several elements that would have allowed a sufficient characterization of its stability. The surveys carried out made it possible to draw conclusions about the type of existing foundations and the state they were in (i.e., the video and a report made by a company specialized in underwater inspection in 1986, and the surveys carried out in 1988 and 1998). There were also technical memories with very detailed information about the bridge characteristics and construction conditions [7].

In the photo of the Hintze Ribeiro Bridge shown in **Figure 2**, dated 1931, the geometry and shape of the cross-section of the riverbed are clearly represented. P2 and P3 would have been the most affected by the flow of the river. It is also clear that protection of P2 and P3 already existed in 1931, as well as the sandbank on the inner edge of the curve where the bridge was located. **Figure 2** also shows that P4 was based on the sandbank, with a level of sand identical to that of P5.

Figure 6.
Final position of the pillar 4 (P4), after the bridge collapse. The arrow indicates the direction of the flow [3, 4].

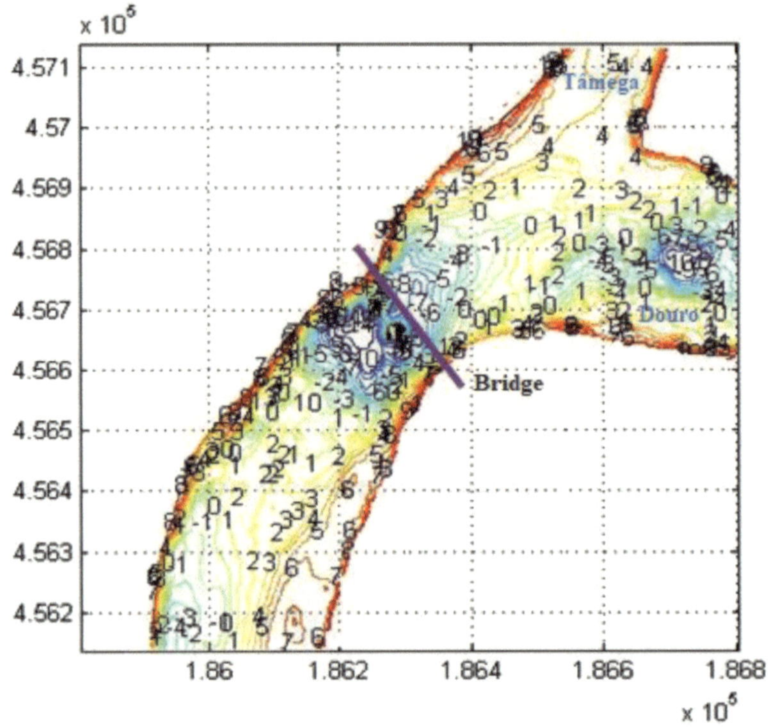

Figure 7.
Douro River survey at entre-os-Rios carried out in 2002. The bluish lines with negative topographic dimensions identify the large existing depressions.

The large volume of sand extracted near the bridge, mainly from the mid-1970s until the construction of Crestuma–Lever Dam and, from there, along the entire length of the reservoir, mainly at various points upstream of the bridge, led to the destruction of the natural sand deposit that existed on the inner bank of the river curve. In addition, the large amounts of extracted sand and the lack of a natural sand supply led to the general lowering of the river bottom and the creation of many depressions along the river, especially in the area where the bridge was located, as documented in **Figure** 7.

The debasement of approximately 13.0 m detected in the period from the mid-1970s to 1989 was mainly due to sand extraction, as this stretch of the Douro River the Douro River is an area of natural deposition, as evidenced by the photo dated 1931 shown in **Figure** 2.

3. The causes of the bridge collapse

3.1 Hydrodynamic causes

By reason of the successive and intense rains recorded in the period between 1 November 2000 and 7 March 2001, the average daily flow of the Douro River remained high for long periods of time, even exceeding the values recorded in previous historical floods. It was also found that the river flow remained very high from 4 March to at least 7 March 2001. Moreover, it is known that floods in alluvial beds, such as the Douro River where the Crestuma–Lever reservoir is located, cause changes in the bottom levels, although only temporary.

From December 2000 to March 2001, the main downgrade may have been caused by high floods (**Figure 5**). However, as this cause was not the only one, it is important to assess whether the bed levels at the beginning of the persistent floods were equivalent to those that existed in previous and equally intense floods, such as those of January 1996 (14 consecutive days), December 1989 (two floods lasting three and four days, respectively), and February 1979 (two floods lasting eight and six days, respectively). If differences can in fact be observed, the total drawdowns (and not just those related to each individual flood), would not have been the same and, as such, the last flood should not be considered as the only decisive factor of the river bottom degradation.

To compute the distribution of the velocity modules to which the P2, P3, P4, and P5 would have been subjected from 5 December 2000 to 4 March 2001, a two-dimensional morphodynamic model (*morphodyn*) in the horizontal plane (2DH) was used [4, 8]. The hydrodynamic module of this computational model solves the Saint-Venant equations, or shallow water equations, which can be deduced by vertically integrating the fundamental equations of fluid mechanics applied to a three-dimensional flow. Fluid incompressibility and hydrostatic pressure are assumed. These equations are written in the following conservative form [4, 8]:

$$
\begin{aligned}
&\frac{\partial H}{\partial t}+\frac{\partial U}{\partial x}+\frac{\partial V}{\partial y}=0 \\
&\frac{\partial U}{\partial t}+\frac{\partial F}{\partial x}+\frac{\partial G}{\partial y}=E_x \\
&\frac{\partial V}{\partial t}+\frac{\partial G}{\partial x}+\frac{\partial S}{\partial y}=E_y
\end{aligned}
\tag{1}
$$

In the system of Eq. (1), all variables are functions of $x, y,$ and t. With $H = h$, the unknowns U, V, F, G, and S are written as $U = hu$, $V = hv$, $F = u^2h + \frac{1}{2}gh^2$, $G = uvh$, and $S = v^2h + \frac{1}{2}gh^2$, where h is the water depth, (u,v) are the depth averaged flow velocity components, and g is the acceleration due to gravity. The terms E_x and E_y in the second and third equations are written:

$$E_x(x,y,t) = -gH\frac{\partial \xi}{\partial x} + fV - \frac{1}{\rho}\tau_{fx} + \frac{\partial}{\partial x}\left(\varepsilon\frac{\partial U}{\partial x}\right) + \frac{\partial}{\partial y}\left(\varepsilon\frac{\partial U}{\partial y}\right) \tag{2}$$

$$E_y(x,y,t) = -gH\frac{\partial \xi}{\partial y} - fU - \frac{1}{\rho}\tau_{fy} + \frac{\partial}{\partial x}\left(\varepsilon\frac{\partial V}{\partial x}\right) + \frac{\partial}{\partial y}\left(\varepsilon\frac{\partial V}{\partial y}\right) \tag{3}$$

In Eqs. (2) and (3), ξ means bed levels, ε is the turbulent diffusion coefficient, and f is the Coriolis parameter, given by $2\Omega sin(\phi)$, where ϕ is the latitude and Ω is the angular velocity of the Earth about its axis, given by $\Omega = \frac{2\pi}{86160}s^{-1}$. The bed shear stresses are computed by the Manning–Strickler formula, which are given by:

$$\frac{\tau_{fx}}{\rho} = \frac{g(u^2+v^2)^{1/2}u}{K^2h^{7/3}} \qquad \frac{\tau_{fy}}{\rho} = \frac{g(u^2+v^2)^{1/2}v}{K^2h^{7/3}} \tag{4}$$

In Eq. (4), $K = 1/n_s$ represents Manning's roughness coefficient. In terms of the equivalent sand grain diameter, n_s is given by $n_s = k_s^{1/6}/21.1, k_s{\approx}2.5d_{50}$.

The system formed by Eqs. (1) is solved numerically using an explicit finite difference method based on the MacCormack's time-splitting scheme [9–11]. In order to apply this method, the Saint-Venant Eqs. (1) are split into two systems of three equations throughout the Ox and Oy directions. The corresponding operators, L_x and L_y, take the following forms (5) and (6):

Operator L_x

$$\begin{aligned}
&\frac{\partial H}{\partial t}+\frac{\partial U}{\partial x}=0\\
&\frac{\partial U}{\partial t}+\frac{\partial F}{\partial x}=-gH\frac{\partial \xi}{\partial x}-\frac{\tau_{fx}}{\rho}+fV+\frac{\partial}{\partial x}\left(\varepsilon\frac{\partial U}{\partial x}\right)\\
&\frac{\partial V}{\partial t}+\frac{\partial G}{\partial x}=\frac{\partial}{\partial x}\left(\varepsilon\frac{\partial V}{\partial x}\right)
\end{aligned} \tag{5}$$

Operator L_y

$$\begin{aligned}
&\frac{\partial H}{\partial t}+\frac{\partial V}{\partial y}=0\\
&\frac{\partial U}{\partial t}+\frac{\partial G}{\partial y}=\frac{\partial}{\partial y}\left(\varepsilon\frac{\partial U}{\partial y}\right)\\
&\frac{\partial V}{\partial t}+\frac{\partial S}{\partial y}=-gH\frac{\partial \xi}{\partial y}-\frac{\tau_{fy}}{\rho}-fU+\frac{\partial}{\partial y}\left(\varepsilon\frac{\partial V}{\partial y}\right)
\end{aligned} \tag{6}$$

Considering the generic variable Q, the solution at time $(n+1)\Delta t$ for the computational point (i, j), is obtained through the following symmetric application (4, 8):

$$Q^{n+1}=L_x\left(\frac{\Delta t}{2}\right)L_y\left(\frac{\Delta t}{2}\right)L_y\left(\frac{\Delta t}{2}\right)L_x\left(\frac{\Delta t}{2}\right)Q^n \tag{7}$$

where each operator, L_x e L_y, is composed of a predictor–corrector sequence and n represents a generic time t. In the four predictor–corrector sequences of application (7), alternatively backward and forward space differences are used. The derivative discretization is performed as follows [4, 8]:

First operator L_x :	Predictor–backward differences
	Corrector–forward differences
First operator L_y :	Predictor–backward differences
	Corrector–forward differences
Second operator L_y :	Predictor–forward differences
	Corrector–backward differences
Second operator L_x :	Predictor–forward differences
	Corrector–backward differences

For the analysis, four hypotheses of initial conditions were considered: Bathymetry of the year 1982, with and without the sand deposit, and the year 2000, with and without the large depressions caused by the dredging that occurred upstream and downstream of the bridge. As the boundary conditions, the discharge flows of the Torrão and Carrapatelo dams were considered, plus 6% of the Carrapatelo flows, for taking into account the intermediate flows of the basin, namely, those of

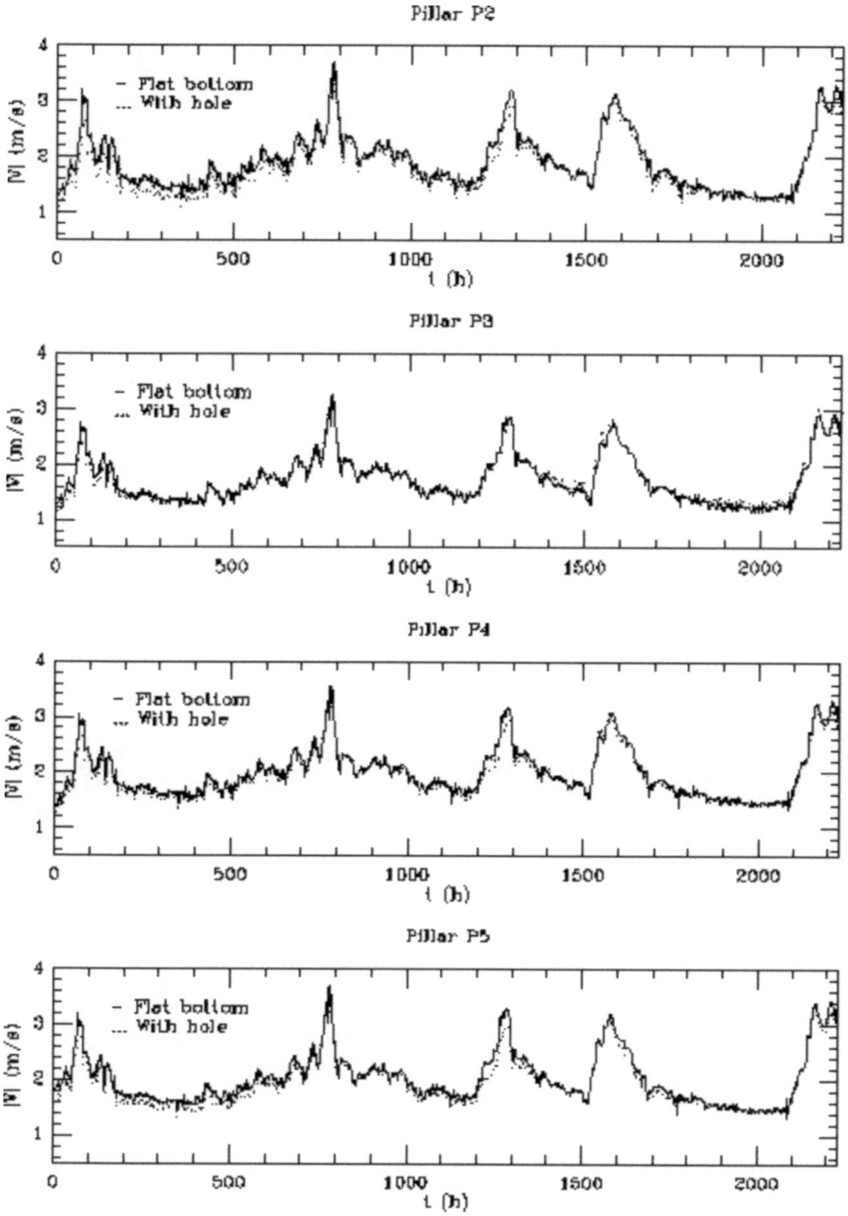

Figure 8.
Velocity module distributions, over time, near P2, P3, P4, and P5, in the period of 5 December 2000 to 4 march 2001.

the Paiva River basin. A grid with cell size of 25 m was used in each simulation. Considering a bathymetry of February 2000, with and without large depressions, the temporal distribution of the velocity modules, i.e., the vertically integrated average velocity, is presented in **Figure 8**, near the P2, P3, P4, and P5 from 5 December 2000 to 4 March 2001.

It is noticed that both P2 and P5 would have been subjected to the greater impacts, with an identical velocity module varying between 1.0 and 3.7 m/s.

Considering the four hypotheses related to the initial conditions, the use of the same model made it possible to estimate the velocity field that would have occurred approximately at the time of the accident (4 March 2001 at approximately 9:00 p.m.).

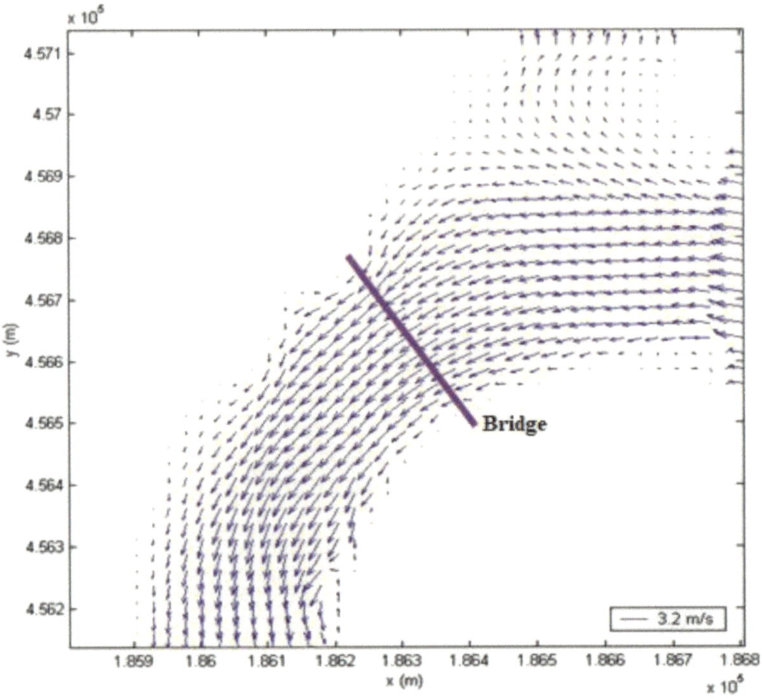

Figure 9.
Velocity field obtained on 4 march 2001 at approximately 12:00 p.m., considering a bathymetry of 2000 without large depressions near the pillars. Note that the flow distribution is 10,000 m^3/s at Carrapatelo dam and 2,000 m^3/s at Torrão dam, therefore with a 5:1 ratio.

Initial Bathymetry	Pillar P2	Pillar P3	Pillar P4	Pillar P5
1982 without sand deposit	3.6	3.2	3.7	3.4
1982 with sand deposit	3.7	3.2	3.6	3.6
2000 without large depression	3.7	3.3	3.6	3.7
2000 with large depression	3.6	3.3	3.5	3.6

Table 2.
Maximum flow velocities near P2–P5 that occurred during the floods from 5 December 2000 to 4 march 2001 (m/s).

Figure 9 shows one of these situations, which corresponds to the velocity field obtained on 4 March 2001 at approximately 12:00 p.m.

The changes in the river bottom observed in the four topographic surveys carried out in 1982, 1989, 2000, and 2001 led to some variations in the velocity values. However, these variations were not very significant, which is explained by the relationship between the changes in bottom levels, in the order of a few meters, and the high flow depths, in the order of 20 m for the flood peaks under analysis.

In contrast, when comparing the velocity fields obtained on 4 March 2001, considering the initial bathymetries (without and with large depressions), no significant differences are observed, although the riverbeds obtained corresponded to different evolutions in the analyzed period: 5 December 2000 to 4 March 2001 (**Figure 8**).

The differences in velocity in the simulations carried out, considering several initial bathymetries of the river, were easily perceived by comparing the evolution over time of the velocity modules near P2, P3, P4, and P5 for the period from 5

December 2000 to 4 March 2001. This analysis made it possible to construct the following **Table 2** with the maximum values of the velocity modules that occurred for the flood peak on 6 January 2001.

Analyzing **Table 2**, it can be concluded that the velocity differences were very small (<10%) and, as such, cannot be considered significant as to their direct effects.

The existence of a large depression in the riverbed generally leads to a decrease in the depth averaged velocity in its vicinity, when compared to the average velocity outside of the depression. However, this behavior occurs simultaneously with changes in the velocity profile, especially close to the edges of the depression, leading to changes in the turbulence pattern.

In the Douro River area, with large floods in a curved stretch and where very strong secondary currents occur, the pattern of turbulence is already very complex, even without large depressions. Although it is not possible to identify in detail the full complexity of the turbulence generated by dredging, it can be said that, although mild, there may have been an increase in turbulence.

In addition, the rockfill around the foundations of P2 and P3 had little effect on the mean flow velocity fields, as they resulted in a very small reduction in the cross-section. In fact, for a 4000 m^2 flood section, the ripraps corresponded to just 140 m^2, which is a reduction of approximately 3.5%. The errors resulting from the reduction in the flooded section were smaller than the errors made in determining the flood beds and would not have had a significant effect on the general pattern. Locally, the rockfill had a protective effect, as it would have reduced the scour process.

The river flows released by the Torrão and Carrapatelo dams are expected to have influenced the velocity field near the bridge. Indeed, the velocity field corresponding to a given total flow should have changed depending on whether the total flow was released only from Carrapatelo Dam, only from Torrão Dam, or in different quantities from each dam. When the major floods occurred, the most likely flow distributions would have been in the order of 10,000 m^3/s at Carrapatelo Dam, while at Torrão Dam, they would not have exceeded 2000 m^3/s, keeping a 5:1 ratio. The numerical computation made it possible to verify that, in most cases, the flow from Torrão Dam was diverted by the flow from Carrapatelo Dam to the right bank of the river, as shown in **Figure 9**.

It should be noted that the discharge from Torrão Dam consists of clear waters, without the sediments that are trapped in the dam, while the floodwaters of the Douro River present high concentrations of sediments. It turns out that these waters would not have mixed immediately, since the flow from Torrão Dam passed close to the right bank of the river. In fact, less water turbidity was observed on this river bank. Thus, the direct influence of the flow from Torrão Dam on P4 was practically nil. It was also observed numerically that this influence was still insignificant for relations in the order of 40% of the flows from Torrão Dam, as happened during some hours in the winter flood of 2000–2001.

In the numerical computations made using the *morphodyn* model (described above), only some influence on P4 for the flows from the Torrão and Carrapatelo dams of the same order of magnitude was noticeable. This is shown in **Figures 10** and **11** for a relationship between the flows discharged by these dams of 0.50 (possible event, although unlikely) and 1.0 (a very rare event, only possible with careless management of the dams).

3.2 Morphodynamic causes

The topographic surveys made it possible to quantify the temporal evolution of the erosive processes in the riverbed of the Douro River, in the cross-section where

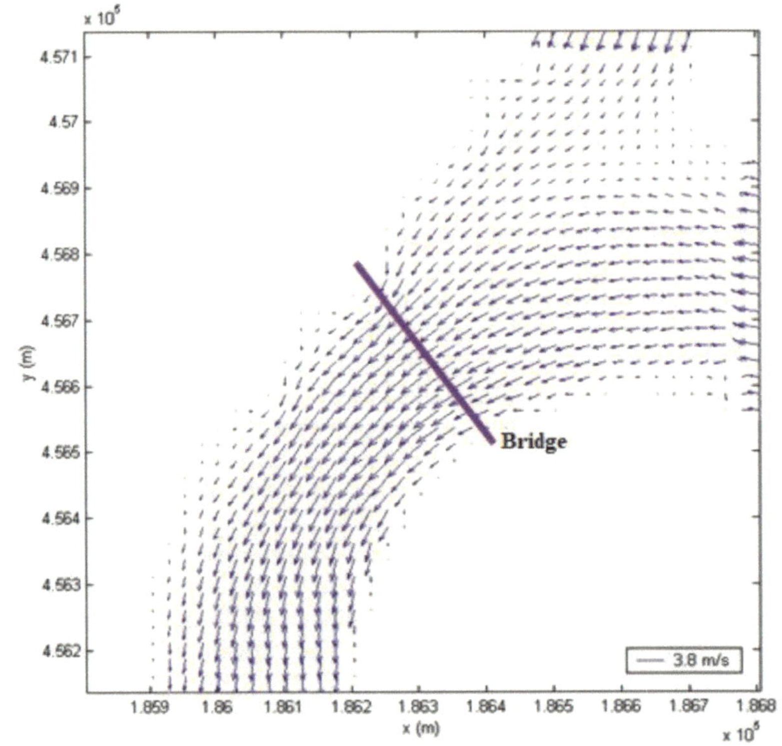

Figure 10.
Velocity field obtained after three days of simulation, considering the bathymetry of the year 2000 without large depressions near the pillars and constant flows of 3000 m³/s in Torrão dam and 6000 m³/s in Carrapatelo dams.

the Hintze Ribeiro Bridge was located, and in the surrounding area. Furthermore, the theoretical analysis of erosive phenomena allowed us to obtain the following conclusions, from the most important to the least important:

1. The general erosion caused by the extraction of sand from the riverbed is an artificial process, which leaves deep scars, and was easily identified in the analyses performed. The comparison of topographic surveys carried out at different times, one before the extractions and the other after the extractions, allowed to quantify the volume differences resulting from the extractions. The sand depression that existed on the curve, clearly visible in old photos (**Figure 2**), as well as on old military maps, disappeared due to the sand extractions that were carried out more intensively from 1975. The large depressions caused by sand extractions were subsequently deformed and partially filled by the intense floods.

2. The scour process around the bridge pillars is partly temporary, with some residual erosion remaining during drought flows (on average).

3. The general erosion that occurs during flooding is temporary, with the previous level of the riverbed recovered some time later; in the Douro River case, it was recovered in weeks.

4. The general drawdown, or degradation, is caused by a tendency of decreasing the supply of sediment. In the Douro River case, this degradation may have

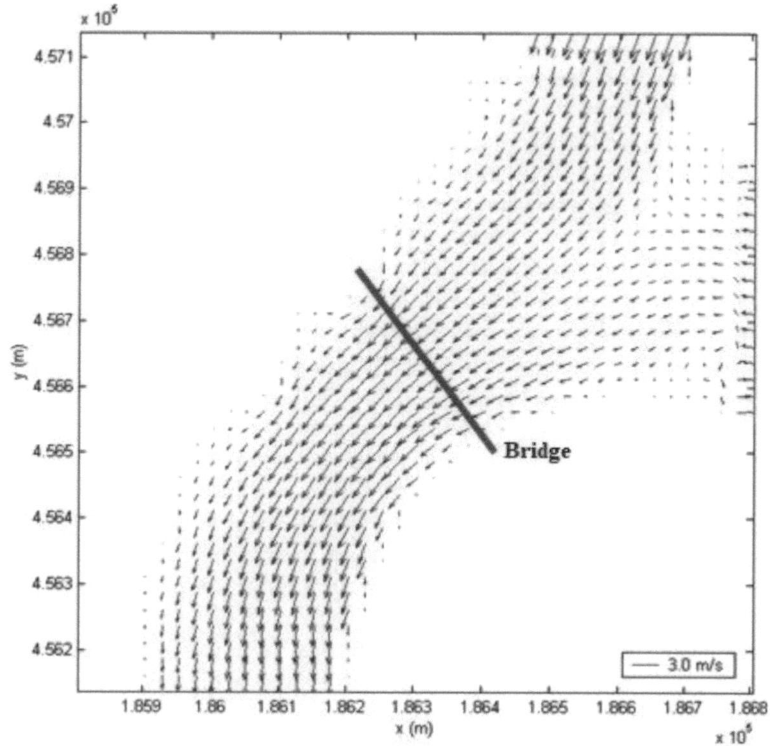

Figure 11.
Velocity field obtained after three days of simulation, considering the bathymetry of the year 2000 without large depressions near the pillars and constant flows of 3000 m^3/s in Torrão dam and 3000 m^3/s in Carrapatelo dam.

increased due to the extraction of sand upstream from the bridge. The large depressions were, in general, fed upstream by the sediments that pass through Carrapatelo Dam, thus retaining the effects of erosion (degradation) downstream, which could be felt up to kilometers away.

The processes of natural erosion begin when the transport capacity of the runoff exceeds the critical stress necessary for the start and transport of sediments by dragging or in suspension. The critical velocity for the beginning of the movement of sandy material is, on the other hand, a function of the granulometry of this material. According to the particle size analysis data of the foundation of P4, the median diameter (d_{50}) of the sediment was in the range of 2–3 mm [3].

The particle size characterization of alluvial riverbeds requires a collection of samples that adequately covers the analyzed area. In this case, there were not enough samples to determine a characteristic value, but a set of values is enough to carry out analysis of sand transport along the river, given the usual errors in other parameters and variables.

According to several equations that appear in the literature, the critical velocity computation, or the initial velocity for the beginning of the movement of sand particles, leads to mean flow velocity values between 0.7 and 0.9 m/s. These critical velocities correspond to river flows between 2200 and 2600 m^3/s [3].

The sediment transport phenomenon is a function of the river flow, the granulometry, and sediment shapes. Therefore, transport occurs at different velocities depending on the size of the particles. In the Douro River case, which may have relatively coarse particles at the bottom, sediment transport from the larger particles, in the order of 6 cm, can occur at velocities above 1.80 m/s, while for the finer particles, around 0.50 mm, sediment transport can occur at velocities around 0.40 m/s [3]. As such, it can be stated that deposition will occur for different particle sizes during flood flows, starting with the coarse ones and ending with the thinner ones. For the coarse sediments, sediment transport can occur for relatively high flows, in the order of 3000 m^3/s.

In the Carrapatelo reservoir, as well as in the reservoirs located upstream, the sediment retention is zero, as was proven by the topographic surveys of all of reservoirs in the Douro River carried out by the Portuguese IPTM (Instituto Portuário e dos Transportes Maritimos) in 2002 (unpublished report). Comparing the longitudinal profile of the Douro River after the great flood of 1962 and the longitudinal profile of 2002 highlighted that there was no silting up in the Douro River dams.

The Torrão Dam reservoir retains practically all of the sediments that arrive there, which was also easy to conclude by the clear color of the water discharged from the dam during the floods, that is, without sediment transport. However, the amount of sandy sediments transported by the Tâmega River is much lower than that transported by the Douro River. In the Douro Basin Hydrographic Plan, a contribution of 37,500 t/year from the Tâmega River is estimated, which corresponds to approximately 7.5% of the contribution from the Douro Basin, which is 500,000 t/year [3].

The decrease in sandy sediment in the Tâmega River is not enough to alter the alluvial behavior of the Douro River. This was confirmed by the topographic surveys carried out at the mouth of the Tâmega River, after the construction of the Torrão Dam, having at that time clarified the great dominance of the Douro River.

This same conclusion was confirmed by the use of the *morphodyn* model, which solved the following sediment conservation Eq. (8), together with the system of hydrodynamic Eqs. (1) (morphodynamic module). Neglecting the contribution of the local variations of sediment concentration [4, 8, 12], this equation is written as:

$$(1-\lambda)\frac{\partial\xi}{\partial t}+\frac{\partial}{\partial x}\left[\langle q_b+q_s\rangle_x-\langle\varepsilon_b|q_b|+\varepsilon_s|q_s|\rangle_x\frac{\partial\xi}{\partial x}\right]$$
$$+\frac{\partial}{\partial y}\left[\langle q_b+q_s\rangle_y-\langle\varepsilon_b|q_b|+\varepsilon_s|q_s|\rangle_y\frac{\partial\xi}{\partial y}\right]=\langle q_{sl}\rangle \tag{8}$$

where $\langle\cdots\rangle$ represents the average quantities, λ is the sediment porosity, q_b and q_s $[m^3s^{-1}/m.l.]$ represent the bed load and suspended load, respectively, per unit width of the watercourse, and q_{sl} $[m^3s^{-1}/m.l./\mathrm{m}.l.]$ represents the input or output sediment transport, per unit width of the main watercourse and per unit width of the tributary.

Such as for solving the system of Eqs. (1), Eq. (8) is also split into the following operators:

Operator L_x

$$(1-\lambda)\frac{\partial\xi}{\partial t}+\frac{\partial R}{\partial x}=\langle q_{sl}\rangle_x \tag{9}$$

Operator L_y

$$(1-\lambda)\frac{\partial\xi}{\partial t}+\frac{\partial S}{\partial y}=\langle q_{sl}\rangle_y \tag{10}$$

where $R=\langle q_b+q_s\rangle_x-\langle\varepsilon_b|q_b|+\varepsilon_s|q_s|\rangle_x\frac{\partial\xi}{\partial x}$ and $S=\langle q_b+q_s\rangle_y-\langle\varepsilon_b|q_b|+\varepsilon_s|q_s|\rangle_y\frac{\partial\xi}{\partial y}$.

Such as in hydrodynamics, the solution at time $(n+1)\Delta t$ for the computational point (i,j), is also obtained through symmetric application (4, 8).

The sediments transported by entrainment (11) and in suspension (12) are written as [4, 8]:

$$\langle q_{b_x} \rangle = \frac{c_f}{g(s-1)} \frac{\varepsilon_b}{tan\varphi} |u|^2 u \qquad \langle q_{b_y} \rangle = \frac{c_f}{g(s-1)} \frac{\varepsilon_b}{tan\varphi} |v|^2 v \tag{11}$$

$$\langle q_{s_x} \rangle = \frac{c_f}{g(s-1)} \frac{\varepsilon_s}{w_s} |u|^3 u \qquad \langle q_{s_y} \rangle = \frac{c_f}{g(s-1)} \frac{\varepsilon_s}{w_s} |v|^3 v \tag{12}$$

where ε_b and ε_s are the efficiencies, with values in the intervals of $0.10 \le \varepsilon_b \le 0.30$ and $0.01 \le \varepsilon_s \le 0.03$, s is the sediment density ($s \approx 2.65$), φ is the internal angle of friction of the sediment, w_s represents the sediment fall velocity, and c_f is a current friction coefficient given by $c_f = \sqrt{0.5 f_c}$, with the bed-roughness coefficient $f_c = 0.06\left(log_{10} \frac{12h}{k_s}\right)^{-2}$, $k_s \approx 2.5 d_{50}$.

Figures 12 and **13** show the numerical evolution of the alluvial riverbed after a three-day simulation, due to the hydrodynamic action of the velocity fields shown in **Figures 10** and **11**, corresponding to the relationships of 0.5 and 1.0 between the flows discharged from Torrão Dam and Carrapatelo Dam, respectively. In these simulations, the initial bathymetry of February 2000 was used, which is basically the one shown in **Figure 7** without the large depressions near the pillars.

Thus, since the transport capacity of the Douro River is much higher than that of the Tâmega River, a decrease in the contribution of sediments from this basin does

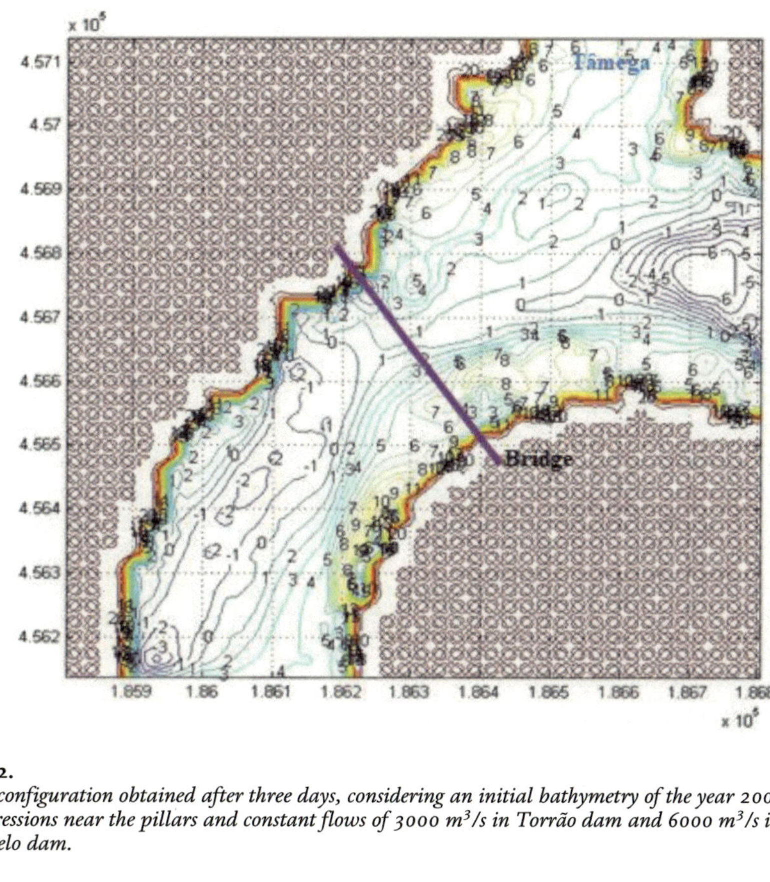

Figure 12.
Riverbed configuration obtained after three days, considering an initial bathymetry of the year 2000 without large depressions near the pillars and constant flows of 3000 m^3/s in Torrão dam and 6000 m^3/s in Carrapatelo dam.

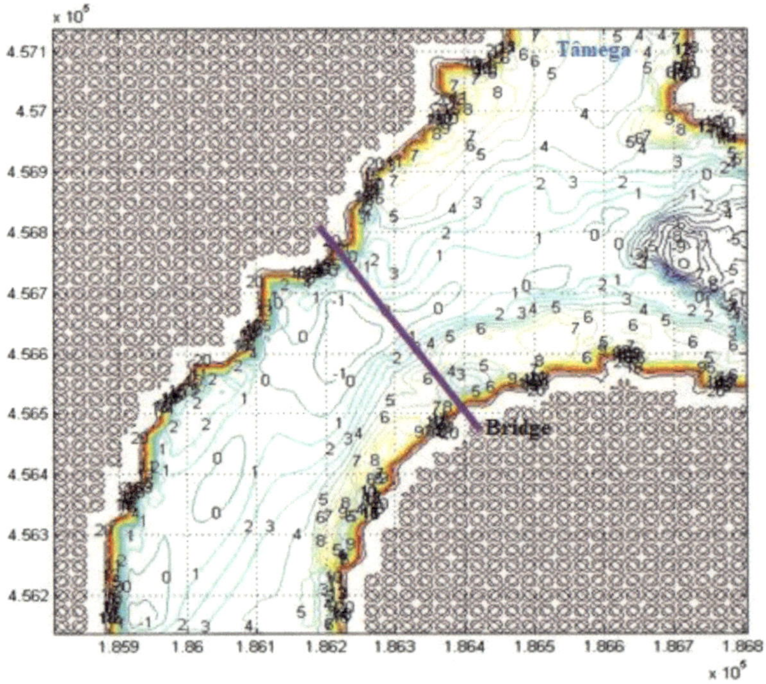

Figure 13.
Riverbed configuration obtained after three days, considering an initial bathymetry of the year 2000 without large depressions near the pillars and constant flows of 3000 m^3/s in Torrão dam and 3000 m^3/s in Carrapatelo dam.

not manifest itself significantly in the general mobility of the alluvial bottom of the Douro River.

4. Discussion and conclusions

The analyses carried out on the various surveys performed since the beginning of the 20th century until the 1970s allowed us to conclude that there were no significant depressions along the stretch of the Douro River upstream of the Hintze Ribeiro Bridge.

Whenever the lowering of alluvial beds is quantified, there must be at least two plots considered: those related to the occurrence of floods (general temporary erosion) and those related to medium-term trends (degradation). The latter is a slower phenomenon, more difficult to reverse, and can have several causes when compared to the drawdowns during floods, namely, imbalances in the natural supply of sediments, upstream reservoir silting, upstream dredging, and riverbed lowering downstream.

Besides the general, slow and fast, and reversible and irreversible lowering, a localized debasement also occurred near P4, as well as close to the other bridge pillars, which were caused by the presence of rigid obstacles.

The large bathymetric depressions identified upstream and next to the bridge pillars, identified in the 1980s and 1990s, correspond to the sum of several causes, namely, the extraction of inert material, the general phenomenon of sediment transport (general erosion), and flow interaction with rigid obstacles, such as pillars (local erosion - scour). These depressions led to a reduction in the natural sediment transport downstream.

With the increase in sand extraction from the 1970s onward, the sand deposit at the left river bank decreased due to direct extraction, having almost disappeared in 1982, as shown in the survey carried out that year (see **Figures 3** and **4**).

Despite the natural complexity of a century-old bridge collapse, the analysis of safety conditions, and considering the hydrodynamic phenomena involved in the Hintze Ribeiro Bridge collapse, it is possible to draw some conclusions. As such, it can be said that:

- In 1931, P4 was implanted in a natural silting area.

- Both P2 and P3 were protected by rockfill before 1931.

- The riverbed lowering in the P4 area was mainly due to sand extraction, as it is a natural silting area.

- The sand levels above the foundation base, before the 2000–2001 floods, were between 7 and 8 m.

- Torrão Dam discharges did not influence the collapse of P4.

- The collapse of the bridge occurred after five consecutive intense floods, which caused P4 to fall due to local scour and general erosion along the river stretch promoted by sand abstraction.

- In the fall, P4 turned and fell upstream, which could only be possible if the paving stones were removed from the foundation base of the pillar by the infra-excavation resulting from scour.

- The riprap placed in P2 and P3 prevented these pillars from collapsing in the 2000–2001 floods and, probably, in other past and equally intense floods.

- The stability of the collapsed P4 could have been guaranteed if a protection solution identical to that found in P2 and P3 had been implemented.

The necessary lessons from the events that led to the Hintze Ribeiro Bridge collapse must be drawn from an in-depth analysis of the causes and consequences of this accident so that catastrophes like this one are not repeated in the future. It is important to emphasize the need to keep hydrographic surveys of riverbeds updated, which can only be done through constant monitoring and effective inspections. Particularly noteworthy is the need to implement restrictions on granting or renewing licenses for extracting sand from alluvial riverbeds. These should be granted only for short periods of time (having the hydrological year as a good reference) and only if certain conditions are met, including:

i. Presentation of a sufficiently detailed topographic survey, covering the areas upstream and downstream of the sediment abstraction stretch and not just the affected area;

ii. Verification that the current levels of the riverbed and that the extracted sand correspond to the values that were authorized;

iii. With the new license, the maximum amounts of sand to be extracted must be fixed, together with the respective abstraction sites.

Author details

José Simão Antunes Do Carmo
FCTUC, Department of Civil Engineering, University of Coimbra, Coimbra, Portugal

*Address all correspondence to: jsacarmo@dec.uc.pt

References

[1] Sousa J.J., Bastos L., Multi-temporal SAR interferometry reveals acceleration of bridge sinking before collapse. Natural Hazards and Earth System Sciences. 2013;13(3):659–667. DOI: 10.5194/nhess-13-659-2013.

[2] Pinto D.F., A Terceira Dimensão. Blogue in http://portugalfotografiaaerea.blogspot.pt/2014/07/entre-os-rios.html 2014 (Accessed October 30, 2020).

[3] Rocha J.S., Antunes Do Carmo J.S., Lemos L.J.L., Silva V.D., Rebelo C.A.S., Bridges built on alluvial beds: The failure of the Hintze Ribeiro Bridge. Recursos Hídricos. 2008;29(2): 41–57 (*in Portuguese*).

[4] Antunes Do Carmo J.S., Natural responses to changes in morphodynamic processes caused by human action in watercourses: A contribution to support management. *International Journal of Disaster Risk Reduction*. 2017;24:109–118. DOI:10.1016/j.ijdrr.2017.05.014.

[5] Teixeira P., A Ponte de Portugal. *Gaiense* journal. 2011 (*in Portuguese*).

[6] Rocha J.S., Alves E., *Plano Específico Preliminar de Extracção de Inertes na Albufeira de Crestuma-Lever, Rio Douro*, LNEC - Laboratório Nacional de Engenharia Civil. 2002, Report 186/02-NHE, July (*in Portuguese*).

[7] Araújo e Silva A.F., Ponte de Entre-os-Rios sobre o Rio Douro, Memória Descritiva. *Revista de Obras Públicas e Minas*, Tomo XIV. 1883:1–24 (*in Portuguese*).

[8] Antunes Do Carmo J.S., Environmental impacts of human action in watercourses. *Natural Hazards and Earth System Sciences Discussions*. 2014; 14:1–32. DOI:10.5194/nhessd-14-1-2014.

[9] MacCormack R.W., Numerical solution of the interaction of a shock wave with a laminar boundary layer. Lecture Notes in Physics. 1971;8:151–163.

[10] Garcia R., Kahawita R.A., Numerical solution of the Saint-Venant equations with the MacCormack finite-difference scheme. International Journal for Numerical Methods in Fluids. 1986;6 (5):259–274.

[11] Antunes Do Carmo J.S., Seabra Santos F.J., Almeida A.B., Numerical solution of the generalized Serre equations with the MacCormack finite-difference scheme. *International Journal for Numerical Methods in Fluids. 1993;*16: 725–738. DOI: 10.1002/fld.1650160805.

[12] Chiang Y.-C., Hsiao S.-S., Lin M.-C., Improved technique for controlling oscillation of coastal morphological modeling system. Journal of Marine Science and Technology. 2011;19(6): 625–633.

Chapter 2

Management of River Basin Physical Assets

Mario Andrés Urrea-Mallebrera, Luis Altarejos-García and Juan Tomás García-Bermejo

Abstract

River basin management involves, among other activities, the operation, maintenance and renewal of existing water and wastewater physical infrastructure assets, as well as the planning, designing, procurement and construction of new water infrastructure assets, in order to provide and secure present and future water demand, and other services, such as flood control and mitigation. Focus is set on increasing demand issues and uncertainty in available resources due to climate change. But there is a challenge also in the management of an aging portfolio of critical infrastructures, including storage, diversion and flood protection facilities, water wells, water conveyance facilities and wastewater treatment plants. Though physical asset management methodologies are well developed and established, such as ISO 55001, their application to infrastructures managed by river basin authorities is not widespread. This chapter presents key components for effective management covering the following aspects: asset monetary valuation; asset condition assessment; estimation of risks linked with asset condition; planning and prioritization of capital and maintenance expenditures; and expected impacts on water tariffs. A raw water distribution system in the Segura River Basin in Spain has been used as case study.

Keywords: water infrastructure, asset management, risk, tariffs

1. Introduction

Since the publication of various international standards for physical asset management, significant and growing research has been carried out in different asset-intensive sectors: industry, water supply and sanitation, water purification, road transport, rail transport and port facilities [1, 2]. Investigations have been aimed at ascertaining the advantages of implementing methodologies in asset management from the point of view of ensuring asset performance, minimizing the risks of potential failures and the consequent effects on users and/or the environment. In particular, asset management has been used by companies in the water sector at different levels of the water supply process, with focus on minimizing failures and reducing potential effects on users, both from a social and economic point of view [3].

The importance of physical assets managed by administrations at different levels has been highlighted [4]. It has been also pointed out the importance of achieving a full understanding of these assets, which requires effective and efficient inventories, not only from a financial and/or accounting point of view, but also considering asset condition and performance. These inventories provide the data needed to carry out risk management of assets in the physical dimension, focused on asset performance and, also in the financial dimension, focused on asset sustainability [5]. It is this importance from an accounting and performance perspective that has led international organizations to focus their analyses on public sector balance sheets, and specifically on some of the lesser known elements, such as non-financial physical assets.

River basin authorities are accountable, among other responsibilities, for providing raw water for irrigation and urban use in a reliable manner, usually managing a complex infrastructure system. This system should be able to deliver access to water with enough quantity and quality levels, under variable hydrologic conditions, while accounting for regulatory and institutional context and operation and maintenance requirements. The system shall also meet current and future, long-term, demand forecasts.

A study of different river basins authorities in Spain covering the 2009–2014 period [6] showed that their degree of financial self-sufficiency was 53% on average value, with values ranging between 24% and 76%. Their economic viability depends on transfers from the government. For instance, in year 2016 a total amount of 225 million euros for capital and 26 million euros for current expenses were transferred. It was found that sources of cash flow imbalances are not only the delays in payments by users of the fees charged by river basin authorities, but also and more important the gap between the needed amount to effectively maintain and renew the asset portfolio and the amount retrieved via tariffs and taxes. There is not an asset management methodology set in place at the river basin authority level and raw water distribution lacks a regulatory framework as far as the management of the physical assets is concerned.

Planning that is currently carried out addresses two aspects:

a. achievement of the objectives of the EU Water Framework Directive, in particular reaching good status of the different water masses in the basin;

b. adequate satisfaction of the demands of the users served.

The diagnosis leads to the conclusion that, since a common methodology for the management of assets does not exist inside river basin authorities, different degrees of deterioration of the hydraulic infrastructure can be found, as a consequence of different efforts in maintenance, conservation and replacement of assets, together with different tax load burdened by the users, which leads to high inequalities in the system.

Therefore, it is needed to improve the management of physical assets operated by river basin authorities, developing methodologies for analyzing the current situation and providing new approaches that allow this very important sector to redirect and redefine the variables used in decision-making by the responsible managers. Portrayal of current situation shows that:

a. infrastructure maintenance and conservation policies have been developed mainly from a corrective point of view, i.e., once the asset have failed, which may not be the optimal strategy, either in terms of asset performance or in terms of the budgetary funds needed to correct it;

b. there are no adequate databases, including sound failure records, to allow performing asset management in accordance with international standards, thus failing to take advantage of the enormous engineering knowledge body that the experts responsible for these organizations have been building up over generations;

c. there is no analysis of the engineering or financial effectiveness of the system so far, particularly when these databases, when they exist, are not properly structured in a homogeneous way.

The investigation carried out in [6] has shown that preventive-predictive maintenance is not widespread across river basin authorities in Spain, there are no protocols for analyzing the current condition of assets and consequently there is no systematic risk analysis of performance associated with the assets. River basin authorities would be, in the best case, at initial stages of implementation of some processes equivalent to good practices, always on an individual basis, not coordinated, and far from requirements of international standards. River basin authorities lack a standardized and homogeneous methodology, which allows for a clear differentiation of budget items in such a way that partial budgets of income and expenses corresponding to the management of hydraulic infrastructures can be differentiated. Consequently, there are no strategic plans for the management of physical assets that involve planning of what needs to be done in the short and medium terms. There is no systematic analysis of actions to be undertaken, how they should be financed, and what impacts may be expected on water tariffs under current legislation.

Therefore, it is considered appropriate, and to a certain extent necessary, to define a common approach for management of assets that may allow at least a future no-deterioration of the hydraulic infrastructure heritage value, redefining budget needs.

2. Management of river basin physical assets

2.1 Overview of water management in Spain at river basin scale

Water management of the 25 Spanish River Basin Districts, which include both inland waters and transitional and coastal waters, is articulated through a continuous adaptive process that is specified through the monitoring of the current hydrological plan of each district, and its revision and updating every six years. This six-year cycle is regulated at different levels by national and Community regulations that make up a basic procedure, significantly common to all the Member States of the European Union, based mainly on the Water Framework Directive. Spanish hydrological planning must make the achievement of environmental objectives for water bodies and associated ecosystems compatible with socio-economic objectives by meeting the demands for different water uses.

The 25 Spanish river basin districts have approved their hydrological or water management plans for the second planning cycle (2015–2021) established by the Water Framework Directive. These are those corresponding to the 11 inter-community districts, whose competence falls to the Central Administration; the Eastern Cantabrian river basin district, with competences shared between the Central Administration and the Autonomous Community of the Basque Country; and 13 intra-community districts: the three with competences in their preparation by the Autonomous Community of Andalusia, the Balearic Islands, Galicia Costa, the River Basin District of Catalonia and the 7 districts of the Canary Islands, whose plans are approved by the Government of the Canary Islands. The 15 river basin districts located in the Iberian Peninsula are shown in **Figure 1**.

Figure 1.
River basins in Spain (only 15 river basins located in the Iberian Peninsula are shown).

The hydrological plans currently in force, 2015–2021, will have to be revised before the end of 2021, giving rise to new plans for the period 2021–2027, which will incorporate, with respect to the current ones, the adjustments that are necessary for their application until their next revision.

The resolution of the discrepancies between the different river basins' plans corresponds to the National Hydrological Plan, which, from a global perspective, must contemplate a harmonious and coordinated use of water resources capable of satisfying the planning objectives in a balanced manner. For this reason, the preparation of the National Hydrological Plan must involve not only the different public administrations, but also the civil society through a broad process of social participation that begins with the development and approval of the River Basin Management Plans. Eventual unbalances between demand and available resources are addressed at the National Hydrological Plan, mainly through inter-basin water transfers.

The first river basin authorities were created in 1926 by Royal Decree Law, being defined in the Water Law as entities of public law with their own legal personality and different from the State, attached for administrative purposes to the Ministry of Agriculture and Fisheries, Food and Environment, through the Directorate General of Water, as an autonomous body with full functional autonomy.

River basin authorities have been operating uninterruptedly since their birth, playing an important role in hydrological planning, resource management and exploitation, protection of the public water domain, concessions of rights of private use of water, water quality control, design and execution of new hydraulic infrastructures, dam safety programs and management of data.

The river basin authorities are public law entities with their own legal personality and have the autonomy to govern and administer by themselves the interests entrusted to them. They can acquire and dispose goods and rights that may constitute their own patrimony. They can contract and bind themselves to exercise, before the Courts, all kinds of actions without any other limitations than those imposed by the law. Their acts and resolutions put an end to the administrative process.

Therefore, river basin authorities have the competence in the management of assets, basically hydraulic infrastructures, which, whether owned by the State or by themselves, constitute an extensive portfolio, with a very important asset valuation, estimated in 5773 millions of euros by year 2015 [6]. It is clear that, proper maintenance of these physical assets is a major challenge for river basin authorities, in order to carry out the functions entrusted by the legislation and including raw water supply network in the territories where they are located.

The main source of funding is through fees, charges, and tariffs. The following items summarize sources of funding:

a. concession canon (for industrial and pasture activities),

b. water regulation canon,

c. wastewater discharge canon,

d. other equity income, through water use and electricity sales fees,

e. funds from the former Ministry of Agriculture, Fisheries, Food and Environment, now MITECO,

f. European Regional Development Fund subsidies,

g. other transfers from the European Union (Life Funds, Feder Funds, etc.).

2.2 Asset management systems

2.2.1 Asset management according to ISO 55000:2014

The ISO 55000:2014 standard provides an overview of asset management, its principles and terminology, and the expected benefits from adopting asset management; defines the factors that influence good asset management, including the nature and purpose of the organization, its financial limitations and regulatory requirements, and the needs and expectations of the organization and its shareholders and highlights the benefits of proper asset management, including improved financial results, information about asset investment decisions, risk management, and improved services and products. Other benefits include the improvement of corporate social responsibility and organization's sustainability, efficiency, and effectiveness.

The standard specifies the foundations on which asset management is based, which includes monetary valuation of assets, planning to achieve the proposed objectives, and leadership and commitment. It is considered essential to achieve the objectives set, for which is necessary to clearly define roles in terms of responsibilities and hierarchies, and to ensure competence, involvement, and training of the staff. Another key aspect is the relationship between individualized assets and their aggregation into systems. Basically, an asset management system is used to direct, coordinate and control asset management activities.

An Asset Management System can be described as a set of interacting and interrelated elements in an organization, whose function is to establish the asset management policy and the asset management objectives, and the processes required to achieve those objectives.

The advantages observed following the implementation of asset management methodologies in organizations include:

a. reduction of operating costs of assets;

b. reduction of capital costs of the investments made;

c. improved performance;

d. reduction of negative impacts on society, mainly associated with health;

e. reduction of operational risk of the assets;

f. minimization of environmental impacts;

g. enhancement of the organization's reputation within society;

h. better organization's performance;

i. reduction of legal and regulatory risks associated with the operation.

2.2.2 Key activities of asset management

Key activities of asset management that in some way should be adapted to the context of river basin authorities are:

a. developing policies, setting the link between the river basin authority strategic plan and the operation and maintenance plans;

b. developing strategies to achieve the objectives set at the organizational level;

c. creating an asset management plan aimed to define over time, what to do and by whom;

d. monitoring the plan, and if necessary, modify strategies in order to achieve the objectives;

e. training of the staff, to always have the best professionals over time in order to achieve the goals set with the least cost;

f. perform risk management, which represents a critical area in the organization;

g. information management, to document and record asset management processes.

2.3 Features of raw water infrastructure assets

It is obvious that both large hydraulic infrastructures of regulation and storage, as well as the distribution networks for water supply and sanitation networks, together with the different facilities of water treatment, such as drinking water treatment plants, purification plants and desalination plants, constitute a wide set of physical assets that can be cataloged as critical [7]. Raw water infrastructures as physical assets present a series of common characteristics:

a. can be subject to monetary valuation, and consequently be represented in the balance sheets;

b. can be inventoried and managed with the corresponding additions, removals and changes;

c. their monetary value depreciates over time;

d. level of performance decreases over time;

e. play a key role with high responsibility in the provision of a critical public service.

It is essential for the implementation of an effective asset management strategy, to understand that these assets do not represent a set of elements that are immovable in time, but are elements that respond to the surrounding conditions in which they perform. Recognizing that assets have a useful life, their proper management results from the combination of financial and engineering strategies, together with a good understanding and management of the risks associated with their performance. Moreover, the evolution of their condition over time is highly correlated with past efforts in operation and maintenance as reflected by budgets. Therefore, asset management aims to meet the challenge of how to do the best in terms of performance, with the lower possible budget, while minimizing the risk.

2.4 Key components of asset management for river basin authorities

2.4.1 Degree of maturity

The first component focuses on the general question of assessing the degree of maturity of the river basin authority organization in their current implementation asset management practices. The concept of maturity for a given organization is understood as the degree of implementation of the methodologies set out in international standards such as PAS 55, ISO 55000:2014, ISO 55001:2014 and ISO 55002:2018. This initial qualification allows the design and application of the proposed management model to be put into context. The management model should ultimately enable the definition of a Strategic Asset Management Plan from the point of view of defining and optimizing the conservation and maintenance budgets and replacement investments to be made, based on a series of objective indicators.

2.4.2 Inventory of assets

The second component would be the inventory of hydraulic infrastructures. The structuring of the inventory should allow, starting from individualized assets, their aggregation into subsystems and subsequent aggregation into a single operating system [5]. Then, a compilation of the available information should be carried out, including new implementation, maintenance, conservation, and replacement actions performed over time, with an indication of the budgets allocated for this purpose.

2.4.3 Asset condition

It is necessary to assess the evolution over time of the monetary value of assets, based on their depreciation over time considering replacement investments and conservation and maintenance actions. It is needed to define indicators of budgetary efficiency and operational sustainability that will make it possible to relate the condition and value of the assets to the conservation, maintenance and replacement actions undertaken in the past. These indicators, in turn, should make it possible to estimate the condition and future value of the assets associated with a given action plan for them [8]. The objective is to describe what has been done in the past in terms of upgrading and maintenance of physical assets, as well as to assess their value and remaining useful life, from a budget point of view.

2.4.4 Risk assessment

A key part of the framework would be to establish a methodology for estimating the risk associated with the different elements that make up an asset, to help prioritize the formulation of action plans for the assets [9]. Prediction of future risk levels after risk mitigation actions have been adopted helps to identify the most efficient measures.

2.4.5 Planning of investments on assets

Based on the above information, different options for investment plans on assets can be made, with the objective of minimizing both risk and asset de-capitalization. For each investment option, an estimate of the evolution of the future monetary asset valuation would be simulated, together with a prediction of the level of future risk associated with the assets, sub-systems and systems according to the possible actions proposed. The potential impact on tariffs and fees associated with each investment intensity would be also simulated.

As a culmination of the above, plans would be defined, including investments and associated budgets for future actions in the short to medium term, typically 6 years, that will make it possible to contribute to the non-deterioration of assets and to the recovery of initial levels of performance, considering both their technical and financial viability, including the estimation of the impact on tariffs over users.

2.4.6 Strategic asset management plan

The next step would be the drafting of a proposal for a Strategic Asset Management Plan, using the above information as background information. A Strategic Asset Management Plan is a document that specifies how the organizational objectives are to be converted into asset management objectives. In addition, it describes the approach the organization takes to comply with these asset management objectives. Aspects covered are scope, definition of stakeholders, setting, asset management policy, asset management objectives, asset status, change management, asset management strategy, performance management, description of asset management system and how the plan would be evaluated.

2.5 Application to a case study

2.5.1 Case study

Part of the assets operated by the Segura River Basin Authority in south-east part of Spain has been used as a case study. The Segura River basin, shown in **Figure 2**, has and extension of 20,200 km^2. It is the driest area of Spain, with average annual rainfall lower than 400 mm, very unevenly distributed both in space and time. Potential evapotranspiration is approximately 990 mm per year. Average runoff is estimated as 43 mm/year, which is the lowest value in Spain. Main river in the basin is the Segura river, with a length of 260 km and annual mean discharge volume of 700 Mm^3. The Segura River Basin has a population of approximately 2,200,000 people. Water resources available are 1566 Mm^3 per year, including

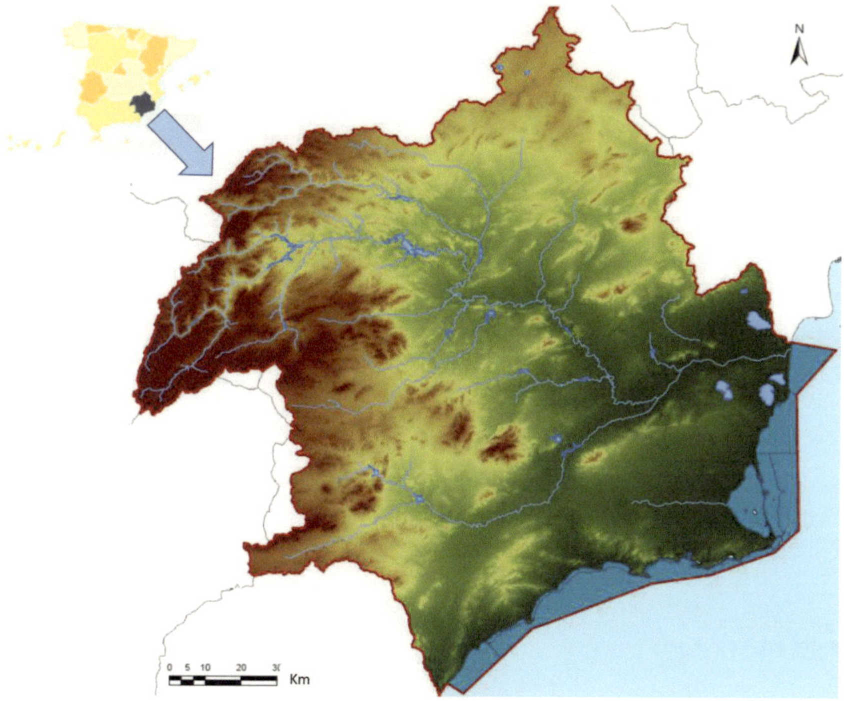

Figure 2.
Segura River basin (source: Segura River basin authority and [10]).

ground water, desalination, reuse, and water transfers from other river basins. On the other hand, water demand is 1843 Mm^3 per year, which results in an average deficit of 277 Mm^3 per year.

The operation of the raw water system in the basin is organized as one single operational system, which is divided into different sub-systems. Each sub-system includes a variety of infrastructure assets: dams, water pumping wells, water mains, pressure pipes, pump stations and wastewater treatment plants. The Segura River Basin system is divided into 5 different sub-systems.

Sub-system 5, for irrigation and domestic water supply, has been used as case study. Sub-system 5 includes 8 assets, as shown in **Figure 3**. Asset n°1 is a reservoir for regulation with capacity of 2.8 Mm^3, with a concrete gravity dam 20 m high. Asset n°2 is a water main with capacity for 10 m^3/s and 128 km long, starting at Asset n°1 and connected with Assets n°3 and n°4. Asset n°3 is a reservoir for regulation with capacity of 1.5 Mm^3, with an embankment dam 32 m high. Asset n°4 is a reservoir for regulation with capacity of 50 Mm^3, with and embankment dam 80 m high. Asset n°5 is a water main with capacity for 30 m^3/s, with a first Section 30 km long before it bifurcates into two branches: one with capacity of 16 m^3/s and 30 km long connected to Asset n°6, and another with capacity of 30 m^3/s and 24 km long connected to Asset n°7. Asset n°6 is a reservoir for regulation with capacity of 15 Mm^3, with an embankment dam 55 m high. Asset n°7 is a reservoir for regulation with a capacity of 220 Mm^3, with an embankment dam 65 m high. Asset n°8 is a water main with maximum capacity of 25 m^3/s and 64 km long.

2.5.2 Inventory of assets

Databases containing information on budgetary investments made since the commission of assets constituting Sub-system 5 have been compiled. Data includes

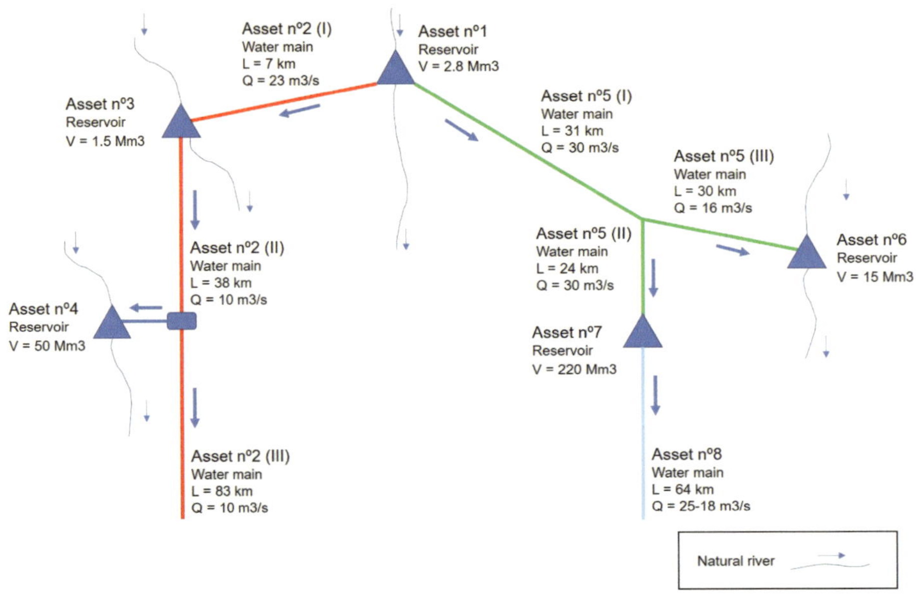

Figure 3.
Schematic of sub-system 5.

actions financed with funds from the Spanish Directorate General for Water, and investments undertaken using the Segura River Basin Authority own funds. This database includes new capital and replacement investments as well as specific maintenance expenditures. Asset valuation sheets include the variables required to undertake monetary valuation over time based upon investments made for new and replacement purposes and those specific to maintenance and preservation. In this case linear depreciation is assumed. Salvage value of assets is expressed as a % of their monetary value at time of commissioning. These percentages vary typically between 80% for civil works and 10% for electrical equipment. Another variable in the database is the useful life of the asset in question. Service life assumed is 50 years for civil works, 25 years for electromechanical equipment and 15 years for electrical equipment.

Monetary valuation of the whole Subsystem 5 at year 2015 is 493.6 M EUR. Value of each asset is shown in **Table 1**.

2.5.3 Asset condition indicators

The index used to evaluate asset condition is the Asset Sustainability Index, ASI(t), which is defined as the ratio between the Cumulative Amount Budgeted for infrastructure maintenance and preservation over the whole life-cycle of the infrastructure, evaluated at time t, CAB(t), and the total Amount Needed to achieve a specific infrastructure condition over the whole life-cycle of the infrastructure, AN, as shown in Eq.(1). The amounts shall include all maintenance actions that contribute to retain the asset in, or to restore the asset to, a certain state, which is the target [10].

$$ASI(t) = \frac{CAB(t)}{AN} \qquad (1)$$

A key aspect for successful use of the index is the accurate assessment of the amount needed, that should consider past experiences of the organization

	Assets in subsystem 5							
	N°1	**N°2**	**N°3**	**N°4**	**N°5**	**N°6**	**N°7**	**N°8**
Monetary value at year 2015 (M EUR)	33.5	128.7	3.7	56.9	40.5	121.4	49.1	59.8

Table 1.
Monetary value of assets of subsystem 5, year 2015 (millions of euros).

together with industry best practices. The amount needed per year is estimated as a fraction of the initial value of the asset. Typical values range from 1% for civil works, 4% for electromechanical equipment, and 8% for access and service roads [10]. Maintenance budget estimated according to these percentages should be understood as an estimate of the minimum maintenance budget which should be provided every year in relation to current replacement costs of the infrastructure, in order to provide a basis for ongoing service delivery. The condition of the infrastructure inferred from the ASI indicator can classified as 'good' (1.0–0.8), 'medium' (0.8–0.6), 'poor' (0.6–0.4) and 'very poor' (0.4–0.2). Values below 0.2 may indicate risk of failure. **Table 2** shows ASI values estimated at year 2015 for the 8 assets included in Sub-system 5.

Analysis of past investments in the full Sub-system 5 for both routine maintenance and replacement actions has been carried out for the period 2008–2015. Average annual value of conservation and maintenance investments is 0.83 M EUR, which is approximately the 0.17% of the assets monetary value. Average annual value of replacement actions is 2.13 M EUR, which represents 0.4% of the monetary value. Altogether, the average annual joint budgetary effort is 2.96 M EUR, which is 0.6% of the monetary value of the full Sub-system 5.

2.5.4 Risk

A risk assessment has been performed, based on the surveys conducted among the management staff and technical staff in charge of assets. A qualitative risk index for the different assets has been determined, on a 0–10 numerical scale. Risk evaluation is done according to the following classification: 'very high' (10–8), 'high' (8–6), 'moderate' (6–4), 'low' (4–2) and 'very low' (2–0). Risk is assessed considering the classical approach of multiplying a qualitative probability of failure index times a qualitative consequence index. Both indexes for probability of failure and consequences are assessed by the river basin authority staff using verbal descriptors of probabilities and verbal descriptor of expected range of consequences. **Table 3** shows Risk Index estimated for year 2015, corresponding to the 8 assets of Subsystem 5.

2.5.5 Planning of investments

Different investment plan alternatives can be prepared based on the results obtained from the operational sustainability indicator, ASI, and the risk level index for 2015, as well as from the analysis of the amounts allocated to replacement investments and to preservation and maintenance. Plans are set for a 6-year time span, corresponding to the period 2016–2021. The 6-year period is distributed in 3 tariff blocks, of 2 years duration each: block n°1 (2016–2017), block n°2 (2018–2019) and block n°3 (2020–2021). Plans include investment efforts of two types: (a) annual investments in asset replacement, and (b) annual budgets for preservation and maintenance.

	Assets in sub-system 5							
	N°1	N°2	N°3	N°4	N°5	N°6	N°7	N°8
ASI (2015)	1.00	0.56	1.00	0.07	1.00	0.29	0.44	0.37
	Good	Poor	Good	—	Good	Very poor	Poor	Very poor

Table 2.
ASI of assets of sub-system 5, year 2015.

	Assets in subsystem 5							
	N°1	N°2	N°3	N°4	N°5	N°6	N°7	N°8
Risk index (2015)	8.05	7.13	6.33	7.06	7.37	6.38	6.33	8.05
	Very high	High	High	High	High	High	High	Very high

Table 3.
Risk index of assets of subsystem 5, year 2015.

	Simulated water tariffs for different user types (EUR/m^3)			
	Past	Investment plan 2016–2021		
	2014–2015	2016–2017	2018–2019	2020–2021
User type 1	0.1219	0.1235	0.1242	0.1249
User type 2	0.1640	0.1652	0.1665	0.1676
User type 3	0.0382	0.0394	0.0396	0.0398
User type 4	0.0579	0.0592	0.0597	0.0601

Table 4.
Simulated water tariffs (in euros per cubic meter).

Considering the monetary valuation obtained for Sub-system 5 at year 2015, the average amounts to be applied on an annual basis of result in 6. 93 M EUR per year, which represents 1.4% of Sub-system 5 monetary value. This amount is divided into asset replacement and renewal, 4.66 M EUR per year, and preservation and maintenance, 2.27 M EUR per year. Given the aging of the various elements, excluding civil works, it would be advisable to carry out a major replacement work over the next six years, 2016–2021, and the following six-year period, 2022–2027.

Assuming that the weight of the investment effort would be fully carried by users, current tariffs should be modified accordingly. Once the structure and the corresponding formulation of the tariff to be applied is known, the simulation of tariff changes is carried out in such a way that only the variables that are affected either by preservation and maintenance actions or by replacement investment actions are modified, keeping constant all the other variables, such as water demand. **Table 4** shows the expected impact on tariffs corresponding to the 4 different kind of users, according to legal framework in the basin, allowing comparison with past tariffs for the period 2014–2015.

With this information, the Strategic Plan for Asset Management should specify the actions to be undertaken during its period of validity, which is intended to be 6 + 6 years. In the first phase, only the actions to be carried out will be detailed, by identifying the asset, type and consequently the sub-system and systems on which the action is to be taken. This is the case of the model tested in Sub-system 5, for that first initial period of 6 years. After the analysis of the tariff impacts, the ASI index and the risk index must be re-evaluated according to the investments made and the new perceptions that the managers responsible for the operation have, in the event that the actions planned for the different scenarios proposed are carried out.

3. Conclusions

The work presented tries to contribute to the improvement of current asset management practices followed by river basin authorities in relation to physical assets of raw water systems. Typical questions that concern decision-makers are:

a. What assets need investment?

b. How much budget should be allocated?

c. When the maintenance or replacement actions should be taken?

d. What is the degree of improvement/minimization of the risk?

e. How does the action program impact on tariffs?

Some gaps and weaknesses have been identified in current practice, including:

a. insufficient disaggregation of budgetary information related to assets;

b. inadequacy of the patrimonial accounting valuation, which makes difficult to support a suitable management of assets;

c. need to improve budget management based on properly constituted cost centers;

d. need for multi-year planning, resulting from a common practice for the different basin organizations across the country.

Decisions on conservation and maintenance of physical assets, as well as decisions on investments for replacement, to be adopted by river basin authorities can be supported by the application of the methodologies and practices described in the present work, which can be considered as a first step in the implementation of the good practices in asset management.

The use of indicators to assess budgetary sustainability based upon historical data of the asset databases managed, as well as the semi-quantitative and qualitative assessment of risk, allows to identify how much budget should be planned and where it should be allocated. Once the critical assets have been identified, different scenarios for replacement investments and maintenance and conservation actions can be proposed, and the impact over future water tariffs can be simulated. The analysis of the impacts on tariffs and the comparison with the existing helps to understand and evaluate the viability and acceptance of these tariffs by the users. The simulation of different scenarios, thanks to the automation of the process of determining the ASI and tariffs, favors the analysis of multiple alternatives that can be adapted to the foreseen financial capacities, or even define the budgetary needs of the period, determined in its case additional financing mechanisms and thus avoiding the progressive decapitalization that has been suffered in the hydraulic asset portfolio.

Author details

Mario Andrés Urrea-Mallebrera[1], Luis Altarejos-García[2*]
and Juan Tomás García-Bermejo[2]

1 Segura River Basin Authority, Murcia, Spain

2 Technical University of Cartagena, Cartagena, Spain

*Address all correspondence to: luis.altarejos@upct.es

References

[1] El-Akruti K, Dwight R, Zhang T. The strategic role of engineering asset management. Int. J. Prod. Econ. 2013;146:227-239.

[2] Too EG. Capabilities for strategic infrastructure asset management [thesis]. Brisbane: Queensland University of Technology; 2009.

[3] Ward B. Integrated asset management systems for water infrastructure [thesis]. Exeter: University of Exeter; 2015.

[4] McDonald RI, Weber K, Padowski J, Flörke M, Schneider C, Green PA, Gleeson T. Water on an urban planet: urbanization and the reach of urban water infrastructure. Glob. Environ. Chang. 2014;27:96-105.

[5] Lee H, Shin H, Rasheed U, Kong M. Establishment of an inventory for the Life Cycle Cost (LCC) analysis of a water supply system. Water. 2017;9:592.

[6] Urrea-Mallebrera MA. Analysis of asset management methodology in hydraulic infrastructures. Methodological proposals for determination of tariffs and taxes under different regulatory frameworks [thesis]. Cartagena: Technical University of Cartagena; 2020.

[7] Zio E. Challenges in the vulnerability and risk analysis of critical infrastructures. Reliab. Eng. Syst. Saf. 2016;52:137-150.

[8] Davis P, Sullivan E, Marlow D, Mamey D. A selection framework for infrastructure condition monitoring technologies in water and wastewater networks. Expert Syst. Appl. 2013;40:1947-1958.

[9] Salman A. Reliability-based management of water distribution networks [thesis]. Montréal: Concordia University; 2011.

[10] Urrea-Mallebrera MA, Altarejos-Garcia L, García-Bermejo JT, Collado-López B. Condition assessment of water infrastructures: application to Segura River Basin (Spain). Water. 2019;11:1169.

Chapter 3

Transboundary River Basin Governance: A Case of the Mekong River Basin

Mak Sithirith

Abstract

Conflict and cooperation are key governance challenges in transboundary river basin governance, especially in the Mekong River Basin. Hydropower dams have been at the center of such a conflict and cooperation that are useful metrics to assess the level and intensity of conflict and cooperation in transboundary river basin governance. This study examines transboundary river basin cooperation in the Mekong through the lens of hydropower dam projects. It uses a literature review and a case study of the Lower Sasan 2 (LS2) Dam to analyze the conflict and cooperation in the Mekong region, from the era of the US influence in the Cold War, the post-Cold War period, and the present-day with the rise of China. It concludes that Mekong river basin cooperation has evolved as a result of external influences and internal competition by riparian states over Mekong resources. The LS2 was identified in 1961 by US-supported hydropower studies and then by the GMS/ADB in 1998, but left unattended until 2007 when Vietnam signed an agreement with Cambodia to undertake a feasibility study in 2008. It took 16 years to get the LS2 built by a Chinese company in 2014 and completed it in 2017. Through the process, the states, powerful external actors, financial institutions, and private sector actors have politicized the LS2 studies, design, and construction. Cambodia, as a weak downstream state, has had to and must continue to position itself strategically in its relationships with these hydro-hegemons to compete for hydropower dam projects and protect its interests. The rise of China has induced the changing relationship between riparian states. Many hydropower dams were built with Chinese funding. Cambodia has also enjoyed its close ties with China, and the building of the LS2 dam by a Chinese company contributes to changing its positions in the Mekong cooperation but suffers environmental and social impacts.

Keywords: transboundary water governance, asymmetric power, hydro-hegemony, hydropower dam, climate change, regional cooperation

1. Introduction

Conflict and cooperation have long been a key governance challenge in transboundary river basin governance. Hydropower dam projects have been at the centre of transboundary river basin cooperation but also the centre of conflicts. These projects have been utilized by riparian states to control water flow and other river resources within their territories, sometimes to the detriment of neighboring states.

Therefore, hydropower dam projects are useful indicators for assessing the level and intensity of conflicts and cooperation in transboundary river basin governance. Thus, the purpose of this study is to examine the transboundary river basin cooperation in the Mekong through the lens of hydropower dam projects. The methodology used consists of a literature review and a case study of the Lower Sesan 2 (LS2) Dam in the Lower Mekong River Basin in Cambodia to explore the dynamics of riparian cooperation and conflict in the Mekong.

2. The Mekong River

The Mekong River is the tenth largest river in the world. The basin of the Mekong River drains a total land area of 795,000 km^2 from China's Qinghai-Tibet Plateau to the Mekong Delta. The Mekong River is the twelfth longest river in the world, flowing approximately 4,909 km through three provinces of China, continuing into Myanmar, Laos, Thailand, Cambodia, and Vietnam before entering into the South China Sea. It originates in the northeastern rim of the Tibetan Plateau situated about 5,000 m above sea level. In China, the Mekong River is called the Lancang River, flowing over 2,200 km. At the end of China's border, the altitude of the Mekong River has dropped to about 400 m above sea level, and it then flows for 2,709 km through Myanmar, Laos, Thailand, Cambodia, and Vietnam [1].

The Sesan, Srepok, and Sekong (3S) rivers form the largest tributaries of the Mekong River, located at the corners of Cambodia, Laos, and Vietnam (see **Figure 1**). It is the second-largest sub-basin of the Mekong River system [2, 3], covering an area of about 78,579 km^2, accounting for about ten percent of the entire Mekong Basin, and spanning the three countries of Laos (29 percent), Cambodia (33 percent), and Vietnam (38 percent) [4]. The Sesan and Srepok rivers originate in the Central Highlands of Vietnam, flowing from Vietnam to Cambodia over a distance of 462 km and 425 km respectively, and meet the Sekong River in Stung Treng Province in Cambodia [3]. The Sekong River originates in the Central Highlands of Vietnam, flowing through Laos and then in Cambodia, where it joins the Mekong River in Stung Treng Province.

The Mekong is home to about 260 million people who are living dependent on the rivers, its water, and other natural resources. About 60 million people live in the Lower Mekong Basin (which does not include China or Myanmar), of which about 3.59 million people live in the 3S basin. Vietnam's population in the 3S basin constitutes 86% of the 3S population. About 60% of the Mekong basin's population relies on agriculture, and 70% of the water diverted from the Mekong each year is used for irrigation, with irrigated land covering 27% of the basin. The Mekong's industrial uses include its potential as a transportation corridor between and inside riparian countries, in particular for the rapidly growing China-ASEAN (Association of Southeast Asian Nations) trade [5].

Capture fisheries make up the largest contribution to the Mekong's fishery, contributing to 2 million tons a year. The 3S River Basin is also rich in fisheries [6]. Fisheries account for nearly 6–9% of Cambodia's GDP and 7% of Laos's GDP. However, the Mekong fishery sectors in Thailand and Vietnam add well over USD 750 million to each of these countries' GDPs per year, although it is less significant to its national economies [7].

The Mekong River Basin has the potential to generate 30,000 megawatts (MW) of hydropower, with only 4,450 exploited as of 2012, of which only 1,600 of that is accounted for the Lower Mekong Basin. The 3S River Basin of the lower Mekong is a potential region for significant hydropower development, with estimated outputs at approximately 6,400 MW [8], mostly in Vietnam and Laos [9]. Hydropower

Figure 1.
Map of the studied areas.

development in the 3S Basin, consisting of individual dams and cascade dams, is rapidly increasing, particularly in Vietnam and Laos, following the pace of economic growth and a need for increased security in the production of electricity.

The LS2 Dam situates at the intersection of two international tributaries of the Mekong River—the Sesan and Srepok River. This chapter explores conflicts and cooperation in an international river basin. First, it examines hydro-hegemony in transboundary river basin management by analyzing the history and politics surrounding the development of the hydropower dams in the Mekong region. Secondly, it looks in depth at the case of the LS2 Dam hydropower project concerning the involvement of the US in Lower Mekong riparian politics and at its importance within the Greater Mekong Sub-region (GMS) and the Cambodia-Laos-Vietnam Development Triangle. Third, it analyzes the role of China in the Mekong and its involvement in hydropower dam development in the Mekong, especially how it is competing with Vietnam to take control of the LS2 Dam. Additionally, the Chapter details the construction of the LS2 Dam by a Chinese developer. It also explores the impacts of hydropower dam construction on local communities and how they respond to Chinese hydropower projects to attempt to protect their livelihoods. Finally, it concludes and makes recommendations regarding how to improve transboundary water governance in the 3S Basin. Further, the findings presented in this manuscript make valuable contributions to the research regarding the impacts of hydro-hegemonies in transboundary river basin governance as well as the growing literature on power and politics in influencing international water relations and shared transboundary discourse.

3. Conceptual framework: conflicts and cooperation in transboundary river basin governance

There are 261 international river basins around the world, where the water and other riparian resources from one river basin is shared by two or more countries. International rivers may be managed for multiple uses such as hydropower development, food production, industrial development, municipal water supplies, recreation, or a combination of these. Different user groups in different countries with different objectives often have difficulty in arriving at a common agreement regarding quantity, quality, and water distribution [10]. The increasing competition for limited freshwater resources along international rivers could escalate interstate conflicts. Thus, the international river basin is a source of a potential interstate conflict, particularly when it comes to negotiating the sharing of water between upstream and downstream states. Water conflicts can be further divided into problems of water quality, water quantity, and ecosystem issues [11, 12]. However, the costs of armed conflict over a shared river far outweigh the benefits of potential victory [13]. So, rather than going to war over water, co-riparian states often choose to cooperate and share river resources and benefits. Globally, Wolf [10] found no water wars within the 261 international river basins across the globe. In the Mekong River Basin, several studies have identified no major conflicts and riparian states within the region cooperate to manage and share the benefits from the Mekong River, despite interstate tensions due to the building of large-scale dams [14, 15]. Nonetheless, the absence of war does not mean the absence of conflict [16]. Indeed, cooperation is reached through a process of dialog, negotiation, discussion, and opposition. Some of these tensions co-exist to the point that they are too complex to be categorized as either conflict or cooperation. These complex tensions within transboundary riparian governance are best understood as water-related interactions rather than as cooperation or conflict [17].

Water-related interactions among many different actors have been driven by power asymmetries that influence decision-making and water allocation among riparians, and thus, it is difficult to represent the variations in these relations over time and through changing economic and political dynamics [18]. Zeitoun and Warner [16] provide a framework to analyze power asymmetry within transboundary river basin cooperation. From the existence of power asymmetries that drive water interactions, it can be seen that an arrangement understood as "cooperative" from a hydro-hegemon's perspective may conceal an inequitable status quo. The weaker states in such an arrangement may have agreed to cooperate for tactical reasons, unrelated to the immediate water issues at hand, and, therefore, the degree to which this arrangement is truly "cooperative" should be questioned [18]. Stable situations of shared control may camouflage unjust outcomes for marginalized stakeholders [16]. The blind pursuit of "cooperation" can, therefore, be as detrimental to the equitable resolution of transboundary water issues as absolute conflict. Furthermore, cooperation between riparian states regarding transboundary water management traditionally has been seen and interpreted as infringing on the sovereignty of riparian states [19]. Hydro-hegemony in transboundary river basin cooperation has often been challenged by counter-hegemony [20].

Zeitoun and Warner [16] describe counter-hegemony as the various strategic responses employed by states sharing the international rivers against perceived negative forms of hydro-hegemony to improve their situation vis-à-vis the hegemon. These strategies could include recourse to ethical arguments and international law, de-securitization, issue linkage, economic development, alternative funding sources, negotiation, and generation of positive-sum outcomes [16]. Dinar [21] introduces side-payment and cost-sharing in international river basin governance as examples of ways to counterbalance the asymmetric geographical relationship between downstream and upstream states. On the other hand, Zawahri [22] points out that third-party involvement in transboundary river basin interactions can leverage the power of riparian states to encourage exploitation of river basin resources, and demonstrates that side-payments from third-party actors such as international organizations, bilateral aid, and non-governmental organizations (NGOs) could reinforce the hydro-hegemon status of riparian states by boosting their capacity for exploitation.

The employment of counter-hegemonic strategies has increased in the past few decades due to the rise of democratization in the post-Cold War era, which has empowered an increasing number of people in different states across the globe to exercise supreme political authority over a geographical region. The post-Cold War era of democratization and globalization has also seen a proliferation of non-state domestic actors influencing national decision-making, including foreign policy. Along these lines, Nicola Pratt [23] has described the rise of civil society as a force for political change, social awareness, and local resistance as a type of counter-hegemony through which challenges to hydropower dam projects by riparian states, including hydro-hegemons, have been made in places such as Thailand and some countries in Europe [24].

Hirsch [25] views the hydro-hegemony in the international shared water as a form of geopolitics. The World's geopolitics has changed after the Cold War, transforming the global geopolitics, particularly the hydro-hegemony, into a more complex set of inter-state relations. The rise of China has resulted in changing inter-state relations and the hydro-hegemonic practices in the Mekong. The region has been re-arranging into overlapping blocs, such as ASEAN, ASEAN Economic Community (AEC), MRC, the GMS, and the Lancang-Mekong Cooperation in the Mekong Region. The economic rise of China and its concomitant increase in demand for energy and water resources [26]. Burgos and Sophal [27] compare the rise of China to the hungry dragon that would need more resources to feed themselves, and so has looked into its neighboring countries to exploit those resources for its domestic needs. Liebman [28] argues that China has trickled down its hegemonies to its neighbors beyond the Mekong river and water resources, but to extract other resources such as mining, oils, timbers, and agriculture, pursuing its interests regardless of how these affect its downstream neighbors. However, these sometimes challenge the existing hydro-hegemonic structures in the Lower Mekong Basin and sometimes leverage the counter-hegemonies.

4. Methodology

This chapter uses the conceptual discussion above, based on the relevant literature, to analyze conflicts and cooperation in the Mekong River Basin. Based on this conceptual approach, it examines how the rise of China, as well as global and regional changes, have affected transboundary river cooperation in the Mekong. In addition to the conceptual framework based on the existing literature, a field study was carried out in two phases. The first phase was conducted between July 2013 and

July 2014, before the construction of the LS2 Dam. The second phase was carried out between January 2016 and May 2017, during the construction of the dam.

Both primary and secondary data were collected to address the research issues and questions presented. Secondary data was collected from various archival sources at the national, provincial, and local levels. Primary data was collected from the study areas through semi-structured interviews with national and community-level stakeholders in addition to focusing group discussions (FGDs).

Fifty semi-structured interviews were conducted in the first phase and 25 semi-structured interviews were conducted in the second phase. Key stakeholders were purposely selected for interviews based on the time and resources available, including NGOs that worked to support dam-affected communities, representatives from the provincial departments of the Ministries of Agriculture, Forestry and Fisheries (MAFF) and Water Resources and Meteorology (MOWRAM) as well as commune council members in Stung Treng Province.

Two FGDs were organized in old Kbal Romeas village—a men's group and a women's group, each participated in by nine villagers. One FGD was conducted in the old Srae Kor Muoy and Pir villages with six women and five men who had not left their villages. The interviews focused on water management, the impacts of dams on rivers and livelihoods, construction of the LS2 Dam (**Figure 1**), compensation, resettlement, the role of NGOs and local government, the intervention of national government as well as support programs for dam-affected communities.

5. Conflicts and cooperation in the lower Mekong River basin

Cooperation in the Mekong River Basin has evolved and has been driven by geography, geopolitics, hydro-hegemons, and national interests. After the independence of Indochina states in the early 50s, Cambodia, Laos, South Vietnam, and Thailand joined the capitalist sphere led by the US, while North Vietnam joined the communist bloc with China, led by the USSR. In the early 60s, the USSR continued to influence China and North Vietnam to form the communist blocs in the Mekong region. These made the US and its allies uneasy about losing other Mekong countries to the communists. To prevent this to happen, the US and its allies made a concerted effort to curtail communist influence in the Mekong region [29–31].

Consequently, the Mekong River was strategically employed by the US to counter Communist influences. The US strategy employed the two fronts of transboundary cooperation regarding riparian management as well as modernization and development initiatives to harness the economic potential of the Mekong and its tributaries. Hydropower dams, especially, were considered symbols of modernization, generating hydroelectricity to fuel economic development in the region as a means of reducing or containing the influence of communism [30, 31]. In 1957, with US support, four countries, Cambodia, Laos, South Vietnam, and Thailand established the Mekong Committee (MC). China was not invited to join the MC, as it was not interesting and the MC was specifically established to counter China's power in the region. At that time, China was also not a member of the UN [32].

Between 1957 and 1970s, the US provided technical and financial support for the MC to establish a planning framework for the construction of a series of large dams in the Mekong Basin [33]. The riparian states in the MC were enthusiastic about the planned development of hydropower dams in the Mekong and 3S Basins. After the establishment in 1957, MC undertook a reconnaissance survey of the major tributaries of the Mekong Lower Basin between 1959 and 1960, and the findings published in a report entitled Comprehensive Reconnaissance Report on the Major Tributaries of the Lower Mekong Basin in September 1961, in which numbers of hydropower development projects were proposed to the MC.

The US-Vietnam war in the 1960s and 70s resulted in all of Indochina becoming part of the communist sphere in 1975. The MC was dismantled in 1975, and the development in the Mekong Region slowed to a crawl. However, in 1978, Laos, Vietnam, and Thailand established the Interim Mekong Committee (IMC) to facilitate regional cooperation without Cambodia due to the Khmer Rouge's refusal to establish regional or international ties. However, the work of the IMC was seriously hampered by the absence of Cambodia. In January 1979, Vietnam invaded Cambodia, overthrew the Khmer Rouge, and installed a new government [34]. With Laos and Cambodia as communist allies, Vietnam became a powerful hegemon in the Mekong region and pushed Cambodia to join the IMC. However, Thailand refused to recognize Cambodia's Vietnam-backed government and therefore denied Cambodia's entry into the IMC as Thailand was concerned that the Communist Indochinese states would collectively oppose Thailand's plans to divert water from the Mekong River [31].

In 1988, Thailand changed its regional policy to transform the Mekong Region from 'battlefields' into 'marketplaces' and worked to speed up the process of both regional reconciliation and enhanced opportunities for hydropower dam development. The Paris Peace Accord in 1991 ended Cambodia's international isolation and Cambodia was admitted to the IMC.

In 1992, the Asian Development Bank (ADB) initiated the Greater Mekong Subregion (GMS) cooperation, involving six Mekong region countries to boost regional cooperation, economic and infrastructure development, cross-border trade as well as encourage economically productive uses of untapped natural resources. Specifically, the GMS aims to promote large infrastructure projects in addition to facilitating connectivity among these states, especially increasing the flow of goods, investment capital, and people across borders. Unlike the MRC, China and Myanmar are also members of the GMS [35, 36]. Since the founding of the GMS, the ADB has implemented more than 100 cooperative projects in infrastructure, energy, telecommunications, trade, tourism, and the environment in its efforts to foster multilateral cooperation and development [26].

In 1995, the UN facilitated the establishment of the Mekong River Commission (MRC) and in April of that year, four lower Mekong countries – Cambodia, Laos, Vietnam, and Thailand – signed the Agreement on the Cooperation for the Sustainable Development of the Mekong River Basin [37]. The Agreement provides a legal mandate for the MRC to coordinate riparian activities in the Lower Mekong Region in developing and managing the Mekong River Basin. In late 1996, after the establishment of the MRC, less populated areas along the tributaries were being looked at for new hydropower projects [38].

Nevertheless, the connections between the US and the ADB/GMS have leveraged the power relations of the states in the Mekong and 3S regions [39]. In return, Vietnam has sought to work with Laos and Cambodia in establishing the Cambodia-Laos-Vietnam Development Triangle (CLV) in 1999 to maintain its influence as well as to compete with China and other non-hydro-hegemons. The Cambodia-Laos-Vietnam Development Triangle Areas (CLV-DTA) was set up by the Prime Ministers of three countries; focusing on security, external affairs, transportation, industry, agriculture, trade and investment, social areas, and environmental protection; covering 13 provinces—5 in Vietnam, 4 in Laos and 4 in Cambodia. The CLV-DTA covers various sectors and priority areas: transport, energy, telecommunications, irrigation and water supply, agriculture, tourism, industry, education and health, environmental protection and land management, security, and defense as well as trade and investment; aiming at promoting economic growth and encouraging the development cooperation between neighboring countries. The 2004 master plan of

CLV-DTA drew on the three countries' plans to develop hydropower and increase power-sharing between its neighbors, stating that the three countries will join their efforts and cooperate in hydropower development. The Vietnamese side can assist the Cambodian and Laos in the triangle area to prepare the technical design for and build small- and medium-scale hydropower projects [40].

After the Cold War ended in 1989, global geopolitics shifted from a "bi-polar" (communist and capitalist) focus toward regionalism. China has begun to revise its foreign policies and institutional arrangements toward those that promote regional economic cooperation [41, 42]. In 1992, China joined the GMS as a member [35, 36]. Furthermore, in March 2016, China founded the Lancang-Mekong Cooperation, involving six other Mekong countries, to form a new cooperation framework. Following that, China has built more than 50 large-scale dams (over 50 MW) in Lower Mekong countries: 30 in Myanmar, 14 in Lao PDR, seven in Cambodia, and three in Vietnam [43]. China has built strong diplomatic and economic ties with Cambodia. These have contributed to the surge in Chinese investment in Cambodia. In 2016 alone, Chinese investment capital in Cambodia amounted to USD1.05 billion mainly in the areas of textiles, hydropower, rice milling, rubber production and processing, mining, and construction. By the end of 2017, China had invested USD12.6 billion in Cambodia. About 50 percent of Cambodian land concessions granted since 1994 (4.6 million hectares) are in the hands of 107 Chinese firms [44]. Some 23 Chinese firms are exploring mineral resources in the Kingdom of Wonder. China invested over USD1.6 billion to build seven hydropower projects between 2005 and 2015 and more are planned for the near future [43]. Five of the Cambodian hydropower dams, with a total capacity of 915 MW, were financed by Chinese investors and built in the coastal regions, including Kamchay, Stung Tatay, Stung Atay, Lower Stung Russey Chrum, Stung Cheay Areng, as well as one in the Mekong region, the LS2 [45–47].

6. Revisiting the Mekong River cooperation through the lower Sesan 2 dam

The conflicts and cooperation in the Mekong River Basin could be revealed through the case of the Lower Sesan 2 (LS2) dam in Cambodia. It politically involves the world power, regional power, sub-regional power, and riparian states in the process of identification, design, planning, construction, and operation of the LS2 Dam over a long period. These processes bind state, private sectors; quasi-state owned corporations, and international financing institutions together in a complex politics of regional cooperation that cut across the 3S River and cultural-ecological landscapes of ethnic groups in the north-eastern region in Cambodia.

6.1 The politics of the lower Sesan 2 dam—The US planted, the GMS planned, and the Chinese harvested the fruit

The LS2 dam was produced out of the US's hydropower planning framework, a US strategy to contain the communist influence in the Mekong Region. It was identified in a reconnaissance survey on the major tributaries of the Mekong Lower Basin in September 1961, four years after the MC established (See **Table 1**). Following that, in the 1970s, the MC reviewed the hydropower projects and identified 180 dams on the tributaries of the Mekong River, including sixteen hydropower development sites on the Se San River Basin—5 projects in Cambodia, 10 in Vietnam, and one international project; and fifteen sites in the Sre Pok River Basin—5 sites in Cambodia, 9 in Vietnam and one international project. The LS2 Project was not included [48].

Project	The Mekong reconnaissance study 1961	Mekong Secretariat 1970	WATCO 1984	SKSSNT (Halcrow) 1998	SWECO 1999
Lower Srepok 2	Indicated	No	No	Yes	n/a
Lower Sesan 2	Indicated	No	No	Yes	n/a
Sesan 4	Yes	Yes (Upper Sesan 4 only	Yes (Upper Sesan 4 only	Yes	Yes
Sesan 3	Yes	Yes (Lower Sesan 3 only)	Yes (Lower Sesan 3 only)	No	Yes
Upper Kontum	Indicated	Yes (Kontum area)	Yes (Kontum)	Yes	Yes
Xe Kaman	No	Yes (Xe Kaman 1,2,3,4)	Yes (Xe Kaman 1,2,3,4)	Yes	n/a
Nam Kong 1	No	Yes (Nam Kong 1,2&3)	Yes (Nam Kong 1,2&3)	Yes	n/a

Source: [51].

Table 1.
Major hydropower projects in the 3S river basin and the identification of the LS2.

However, Cambodia did not support the US plans to construct a series of large dams in the Mekong and so, refused to accept it's given the assertion of US influence and dominance in the planning process. In 1965, Cambodia ended its relationship with the US. In return, the US supported the military coup in Cambodia in the 1970s, but the country fell into the Indochina War led to the Khmer Rouge occupation in 1975 and the killing field between 1975 and 1978. In 1979, Vietnam overthrew the Khmer Rouge and controlled Cambodia until the 1989s. The hydropower plans, including the LS2 Project, were abandoned as a result [48].

The UN-supported election in 1993 resulted in opening up Cambodia to regional and international integrations. In 1992, the GMS was established and Cambodia joined it as a member. In the early 1990s, the GMS funded two major preliminary technical and economic feasibility studies for hydropower projects in the region. The first study, undertaken in 1994, provided an initial inventory of potential hydropower projects in the Mekong region. The second study, begun in 1998, specifically focused on the feasibility of hydropower projects in the Sekong, Sesan, and Nam Thoeun River Basins (SKSSNT). The SKSSNT study was completed in two phases and analyzed 37 potential hydropower sites in the three basins. The 11 highest-ranked projects were then selected (See **Table 1**),[1] from which six dams, were selected for further study: the Sesan 3 and Thoung Kontum in Vietnam, the Lower Srepok 2 and Sesan 4 in Cambodia, and the Xe Kaman 3 and Nam Kong 1 in Lao PDR.[2] These proposed projects were formally prioritized largely because of their high potential for electricity generation and because their economic and financial benefits were more favorable than other proposed projects [49, 50].

[1] WATCO is a Dutch Consulting group, who specialized in hydropower development and water resources management.

[2] SWECO is a Sweden's leading consulting engineering company and the group employs 2,300 people with 40 offices in Sweden and 10 others worldwide. The SWECO Group covers a wide range of work in addition to hydropower

The Lower Sesan 2 and Lower Srepok 2 projects in Cambodia were not originally included in the list of proposed projects (**Table 2**) for economic reasons. A subsequent listing politically moved these schemes to the 7th and 8th positions, respectively. Large differences in feasibility and expected economic benefits existed between the projects on the list. The addition of two Cambodian projects to the list was motivated by the fact that projections were almost profitable and that its economic standing was only slightly below the other 11 projects. In pushing the idea of two sites in Cambodia, the study suggested an alternative dam site for the Lower Sesan 2; eventually, that proposal proved uncompetitive and was ranked with lower economic value than the existing projects. Despite this, the original Lower Sesan 2 still appears among the six priority schemes. Reportedly, the addition of the Cambodian projects to the priority list was due to political pressure rather than economic feasibility. Two different locations were proposed for the Lower Sesan 2 [35, 50], although the economic feasibility of the remaining proposal has also been questioned, and it is still included as a priority project in the SKSSNT study, possibly due to the pressure to include two priority dams for each of three countries.

Vietnam takes the GMS/ADB hydropower plan forward. During the second Summit of CLV-DTA in 2003, Vietnam signed the MOU on feasibility studies for three hydropower plants of Lower Se San 1/Se San 5 with a capacity of 90 MW

Project	Economic preparedness, local planning, Score Maximum 52%	Social Score Maximum 25%	Environmental score maximum	Total Score Maximum 100%	Comment
Sesan 3	40.7	23.8	16.6	81	
Xe Katam	33.5	24.8	21.6	79.8	MOU
Sesan 4	37.2	16.5	17.7	71.5	
Houay L.G	29.1	24.3	14.3	67.6	
Upper Kontum	35.1	16.5	15.9	67.4	
Nam Thoeun 1	41.1	15.8	10.6	67.4	MOU
Lower Srepok 2	21.1	23.8	16.8	61.5	
Lower Sesan 2	19.9	20.5	16.3	56.7	
Xe Kam 3	23.5	11.5	20.9	55.9	
Nam Kong 1	21	16	15.9	52.9	
Xe Kam 1	29.3	13.3	9.4	52	MOU
Se Kong 4	27.3	14	7.6	48.9	
Se Kong 5	21.8	12.3	10.8	44.8	

Source: [51].
MOU = Memorandum of Understanding.

Table 2.
Hydropower project in the Sekong, Sesan, and Nam Thoeun River basins.

located on the border of the two countries, Lower Se San 2 with an electricity generating capacity of 400 MW in the lower basin of Se San river and Stung Treng hydropower plant with capacity of 980 MW. Vietnam agreed with Cambodia to build the LS2 Project during the five years from 2012–2017 in addition to completing the Lower Sesan 1/5 (LS1/5), LS2, LS3, Prek Leang 1, and Prek Leang 2 projects between 2015 and 2020 [2, 40]. As part of the CLV, Electricité du Viet Nam (EVN), Vietnam's electric power utility company, signed a memorandum of understanding with the Cambodian government to connect the LS2 and LS1/5 projects to the Cambodian and Vietnamese power grids [52]. The agreement between Cambodia and Vietnam on the LS2 project said that about 50 percent of electricity generated from the LS2 project would be sold in Cambodia and the remaining would be exported to Vietnam, as the power grid in Cambodian beyond the northeastern provinces remains undeveloped and electricity demand in northeast Cambodia is far less than what the LS2 could produce [44].

The LS2 Project was finally selected by China among seven hydropower projects to be built in the Mekong River in Cambodia. China's investment in the LS2 dam in Cambodia serves as a placeholder for larger political and economic issues that are not necessarily energy security or profit-driven. The individual projects, the LS2 Project, do not necessarily need to be profitable to be approved, but rather contribute to magnificent relationship goals between Cambodia and China and their political and economic strategies [43]. On the one hand, Cambodia occupies a unique geographical location in the Mekong and Southeast Asia. Through this relationship with Cambodia, it is strategically important for China to exert greater influence in Southeast Asia and to counterbalance the power of the US in the region. The Cambodian port city of Sihanoukville is a precious "pearl" among China's "string of pearls" and access to these ports provides an excellent base for projecting maritime power into the Gulf of Thailand and the Straits of Malacca [53].

6.2 The conflict and cooperation between China and Vietnam over the building of the lower Sesan 2 (LS2) dam

The building of the LS2 project, though planned, did not happen between the late 90s and early 2000s due to a lack of funding. Vietnam is concerned about China's trickle-down hegemony into the LS2 project and the rest of the 3S region and grabs the economic opportunities to tackle the LS2 project, fulfilling its CLV panning framework. On 15 June 2007, the Royal Government of Cambodia (RGC), represented by the Ministry of Mines and Energy (MME), granted permission for Electricité du Viet Nam (EVN) to conduct a detailed feasibility study for the LS2 project. In 2008, EVN hired Vietnamese and Cambodian consulting companies to conduct feasibility studies, including social and environmental impact assessments [54, 55].

The feasibility study was completed in late 2008. Many concerns were raised about the possible negative impacts of the LS2 project and so, Cambodia's Ministry of Environment delayed approving the project, but requested more studies, particularly on fisheries. In response, the Prime Minister of Vietnam, Nguyễn Tấn Dũng, publicly requested in November 2010 that Prime Minister Hun Sen of Cambodia to speed up the project's approval. The approval came shortly after that request by RGC. As a result, Vietnam's Ministry of Planning and Investment licensed EVN in January 2011 to make a USD816 million investment in the project. The LS2 project was a joint venture between an EVN subsidiary, the EVN International Joint Stock Company, and Cambodia's Royal Group, which together incorporated the Cambodia–Vietnam Hydropower Company as the project owner.

In return, Vietnam was concerned about the high cost of building the LS2 project and so, the EVN looked for a joint venture with private investors. Finally, in

January 2011, the Royal Group of Cambodia agreed to shoulder 49 percent of the project cost, although technically, it has no previous experience in energy development. But it was rumored that Prime Minister Hun Sen personally asked the Royal Group to invest in the project as it had close political ties to Hun Sen. Nevertheless, the project was also economically attractive to the Royal Group as the LS2 site was covered by huge forest areas, which had the potential to generate millions of dollars in timber sales. However, the EVN wanted to withdraw completely from the project, which they finally took about two months before the approval of the project on November 2, 2012, by the Cabinet of the Royal Government of Cambodia, despite the funding commitment of about 50% from the Royal Group.

The uncertainty of the EVN opens up the opportunity for China's Huaneng Group, whose subsidiary Hydrolancang International Energy, to grab the economic opportunity to invest in the LS2 dam and eventually replaced the EVN in the joint venture. The entry of the Chinese into the project meant that Chinese banks provided the capital needed to proceed with the project. The Council of Ministers of the Kingdom of Cambodia approved the LS2 project with a capacity of 400 megawatts (MW), to be built by China's Huaneng Group and the Royal Group on 26 November 2012 [56].

In late November 2012, Hydrolancang International Energy Co. (HIE), a subsidiary of China Huaneng Group signed a memorandum of understanding with the Royal Group for an "initial two-year cash injection" into the LS2 project [57]. Since that time, the project developer has been Hydropower Lower Sesan 2, a joint venture between the Royal Group and HIE, which together own 90 percent of the stakes. The LS2 dam has a 45-year concession period that includes five years for construction. The new arrangement led to a revision regarding sharing investment costs. The joint venture directly financed 30 percent of the project with the remaining 70 percent financed by an undisclosed bank loan. HIE provided the largest share (51 percent) of the project's costs, while the Royal Group reduced its share from 49 percent to 39 percent. The EVN eventually decided to purchase the remaining 10 percent of shares. Finally, the joint venture created the Hydropower Lower Sesan 2 Co. Ltd. to construct the LS2 Dam [58].

The dam was built by Hydro Power Lower Sesan 2 Co. Ltd. at an estimated cost of USD816 million [59]. The LS2 Dam is located near the confluence of the Sesan and Srepok rivers, about 1.5 km downstream from where the Sesan meets the Srepok and 25 km from the confluence with the Mekong mainstream. Clearing of the reservoir area for the LS2 dam began in March 2013. The resettlement and compensation plan was published in January 2014. The construction of the LS2 dam began in February 2014 and was completed in 2017. Prime Minister Hun Sen presided over the ceremonial opening of the LS2 dam on 25 September 2017. At full capacity, the LS2 dam will generate 400 MW of electricity. Moreover, it will reduce the country's reliance on imported electricity from Thailand, Laos, and Vietnam, lower electricity tariffs, and stimulate industrial development. The LS2 dam is the first large dam on the Cambodian section of the 3S rivers [55].

6.3 Ecological, biophysical and environmental impacts

The landscapes of the 3S Rivers consist of biophysical and spiritual landscapes. The biophysical landscape comprises of rich river ecology, the Sesan, and Srepok Rivers, with water flow and fish in rivers. Along the riverbank, forests cover huge areas and communities settled on the riverbanks. Forest and rivers are homes to terrestrial wildlife and aquatic animals. The area is rich in biodiversity. However, villagers believe that forests and rivers are homes to spirits known as 'neak ta' [60]. Four different types of spiritual landscapes have existed in the areas—(1) the land spirit (neak ta), (2) the forest spirit (prey neak ta); and (3) water/river spirit (neakta krahamkor) and (4) ancestral burial grounds. Villages argue that these spirits protect them, bring them happiness and harmonies, and provide them with good businesses and agricultural activities, and protect them when they travel on the river. Villagers pay respects to their respective spirits day, night, and year.

The LS2 dam removes the river landscape and replaces it with a concrete structure, blocking the river flow and fish migration and dividing the river system into the upstream and the downstream areas of the dam. The upstream area creates a huge reservoir, covering an area of 340 km^2 and the downstream area is characterized by the increased water flow throughout the year. The construction of the LS2 dam led to the clearance of 33,564 ha of forest and destruction of biodiversity, of which about 350 ha were evergreen forest, some 5073 ha semi-evergreen forest, and 27,711 ha deciduous forest. Matthews and Motta [43] state: "in the Lower Sesan 2, where reservoir clearing began inside the concession area before dam designs were even approved. The tree clearing at the Lower Sesan 2 was also illegal as it extended well beyond the borders defined in the concession agreement. Many speculate that the central government and the State-Owned Enterprises (SOEs) are signing dam concession agreements to access the valuable timber reserves". On 6th June 2016, an article in the Cambodia Daily calls the LS2 dam a means of laundering illegal timber.

Some 329 fish species used to live in the 3S river systems −133 species reside in the Sesan river, 213 species in the Sekong River, and 240 species in Srepok River [61]. At least 41 migratory fish species are commonly caught by fishermen in the Sesan River, and these migratory species represent 60% of the fishermen's total catch. The Yali Dam in Vietnam alone would reduce fish populations and fish yields by 70% [62]. The LS2 dam has modified the river ecology, disconnecting the upstream with the downstream, destroying the fish habitats and migratory routes of fish, and changing river flow between the dry and wet season. The scientist predicts that the LS2 dam is likely to have the most significant impact on fish biomass levels, which are expected to drop by 9.3% basin-wide, amounting to approximately 200,000 tons of fish each year and further affecting fisheries productivity levels in the Sesan and Srepok river basins [63].

In Srae Kor commune, about 2031 ha of community forestry established in 2013 by the Ministry of Agriculture, Forestry and Fisheries along the Sesan River were cleared by the Hydro Power Lower Sesan 2 Co. Ltd. Inside the area, about 280 ha of prey neak ta or 'forest spirit' that were protected and sacred to villagers for a long time were destroyed on short notice in late 2014 and early 2015. Four burial grounds of about 50 ha each, two located along the Sesan and the other two about 300 m from the river, were removed by the company, which paid compensation of USD150–160 for each grave. Furthermore, about 1248 ha of 'wet ricefield' and 57 ha of upland ricefields were lost to dam construction.

In Kbal Romeas commune, the government granted an economic land concession to Anmady Investment Group for planting rubber, covering about 3000 ha in the north-east part of the commune, denying villagers access to forest resources and agriculture. Furthermore, the Hydro Power Lower Sesan 2 Co. cleared the lands used by villagers for wet rice, covering 620 ha. In Kbal Romeas village, clearance by the company destroyed 'three burial places' covering about 60 ha, 348 ha of the forest spirits, 358 ha of upland fields, a health center, and 232 ha of rice paddies. The compensation covered only the burial land, rice field, and housing.

Neakta Krahamkor is a water/river spirit, the ancient spirit with a red neck, who protects the river and the villagers who depend on it. Villagers gather annually and begin a traditional ritual, offering incense, a bowl of rice, and a pig's head, requesting their pleas and protection. This practice has happened since the early 1900s when a Chinese merchant was rewarded with a successful journey up the river when he paid his respects to the spirit with the red neck. At present, this site has been lost to LS2, an indication that the life of people will be ruined as the protector has disappeared.

6.4 Local resistance to the LS2 dam

Six villages, namely Srae Kor Muoy, Pir, Kbal Romeas, Srae Sranok, and Phluk, were submerged in the reservoir. About 846 families were directly affected by the LS2 dam and a further 15 villages downstream, home to 3,794 families, and a population of 19,066 were indirectly affected. About 70 other villages situated along the Sesan and Srepok upstream of the dam, home to 6,387 households and a total population of 32,864, were also expected to be negatively affected by the project [55].

The affected populations were relocated to four different resettlement areas, covering 4,000 ha, organized by the Project Owners. The project owners proposed the resettlement sites and the provincial government and the relocated chose the locations that best suited them. Villagers from Chrab, Srae Sranok, and Kbal Romeas chose to resettle along National Road No. 78, known locally as the ASEAN Road, linking Stung Treng and Ratanakkiri provinces. However, villagers from Srae Kor Muoy and Pir chose to resettle in the northern part of the Sesan, about 3 km from the river.

The compensation took two forms: (1) a land and housing package and (2) cash. The cash payments for loss of land were USD500 per hectare for the lowland paddy field, USD740 per hectare for garden land, and USD230 per hectare for fallow swidden [55]. The 12 families from Phluk chose the cash package [64]. As part of this package, compensation for fruit trees was very low. For a banana tree grove, for instance, compensation was only USD6. Local people said they could sell the fruit from a single banana tree for much more over the course of a year than the compensation that was provided for an entire grove.

The affected households were eligible for the land and housing package, which was to provide each displaced household with: (1) 5 ha of farmland (2) land for the housing of 1000 m^2, and (3) a house of 80 m^2. For housing, the relocated had two options: a company-built house or a lump-sum cash payment of USD 6,000 to build their own house. Two types of houses were offered – wooden and concrete. About 50 percent of resettled families opted for the wooden houses built by the company, 18 percent chose concrete houses, 20 percent chose to build their own houses and eight percent received only plots of land for housing without a house. They also received a financial package for livelihood restoration during the transition period. However, the lowland paddy soil lost in places such as Kbal Romeas and Srae Kor was of very good quality while the replacement land is not nearly as good and even this low-quality land is available only in limited amounts, often scattered over wide areas. The compensation for lost agricultural land was inappropriate and inadequate.

Interviews with many of them indicated they did not want to move, but the project owners and government required them to relocate. At the same of the studies, about 85 percent of the affected families had relocated to new resettlement sites, but about 15 percent refused to leave their villages. All villagers from Srae Sranok and Chrab villages were completely relocated to the resettlement villages organized by the project, while 126 families from Srae Kor Muoy, Pir, and Kbal Romeas refused to move and decided to stay in their houses and land in the old villages. Unlike the other displaced villagers, 12 families from Phluk were completely removed from their areas to make way for the construction of the LS2 dam(**Table 3**). Adding to the problem, the original plan was changed from offering five different house sizes or types to just one standard-sized house. Thus, housing compensation failed to take into account the size of the household and the number of families in a household.

Village	No. resettled families	No. agreed to relocate	%	No. refused to relocate	%
Phluk	12	12	100	0	0.00
Chrab	47	47	100	0	0.00
Srae Sranok	168	168	100	0	0.00
Kbal Romeas	142	112	78.87	30	21.13
Srae Kor 1	245	202	82.45	43	17.55
Srae Kor 2	232	179	77.16	53	22.84
Total	846	720	85.11	126	14.89

Table 3.
Number of families accepting and refusing compensation.

The reasons that families agreed to relocate included fear of negative project impacts and, importantly, promises of compensation. However, they indicated that relocation was a "no choice" option for them. Mr. Chhang Chhoeun and his wife, Ms. Chrab Veth, with four children, one married, said that he agreed to relocate to a new village and received a wooden house with a land area of 20 x 50 m, and 5 ha of agricultural land. For him and his family, this was acceptable as it is better than what he had in his old village of Kbal Romeas, where he had only 4 ha of farmland and a tiny wooden house for the entire family, including the married daughter's family. In the new village, his married daughter has her own new house and that puts his family in a better position. Chheum Kea, a councilor of Kbal Romeas commune, decided to relocate to the resettlement village to avoid being seen as against the government project. His house in old Kbal Romeas village was about 99 m^2 and his kitchen was 30 m^2. He has three married children. In the new village, he received a two-roomed house of only 80 m^2, which is not big enough for all his married children. His married children did not receive their own houses because they married after the project started. Thus, he was not happy but had to move anyway. These are just two of many stories of relocation bringing both satisfaction and dissatisfaction, although many did not want to move.

Some families refused to relocate because they had built a good house in a good location in the old village, with access to rivers for water, forests for non-timber forest products, and personal religious sites, including the tombs of their parents and grandparents as well as shrines of the forest spirit that they believe provides security for their lives. The new villages cannot guarantee this same quality of life, as many project-constructed houses in the resettlement locations are smaller and made of low-quality wood. Also, many of the new sites lack access to water and drinking water because many of the hand pumps built by the companies did not work and the water quality was not acceptable. More importantly to the villagers, they would be separated from the tombs of their ancestors and the forest spirit.

The livelihoods of ethnic peoples from Srae Kor, Kbal Romeas, Phluk, and Srae Sranok are very much dependent on rivers, forest, non-timber forest products, livestock raising, and agriculture. The Sesan and Srepok rivers provide them a moderate lifestyle: river water is used for both drinking and agriculture and fish, rice, and aquatic vegetables are the main foods. The forest has for generations provided villagers with a source of food, energy, materials, and spiritual life. Land along the rivers and surrounding the villages is cultivated and rice is the main crop. According to the villagers, the rivers, water, forest, and agricultural lands are home to spirits with whom the villagers have lived well and to whom they pay respect through traditional ritual practices. Livestock raising is not only for income and food but social and traditional rituals and practices as well. The villagers in these areas believe that any action leading to damaging rivers, forests or land angers the spirits and without a warm relationship with these spirits, they cannot look forward to a long, happy and healthy life.

Relocation to a new area means that various ethnic groups will have to adopt a new lifestyle based on markets. From a cashless tradition, ethnic groups of Lao, Phnong, and Kreung must switch to the common livelihood of buying and selling. Free collection of resources from rivers, forests, and land will be replaced by paying to obtain them. This is evident in the resettlement sites, where the relocated households now have to buy water, fish, and meat for their families, unlike before in their previous villages. Some of the relocated households have used some of the cash compensation to buy motorcycles, TVs, and other items, such as phones. In the long run, they may face a big dilemma as the cash dries up. Shortage of water, extra expenses, limited income-earning opportunities, and changing cultural practices in the new villages worry the relocated immensely. They feel uncertain about their lives and cannot predict what might happen next. They are calling on the government and the hydropower company to address these issues, but the authorities have been slow to respond.

7. Discussion and conclusion

Cooperation among riparian states in the Mekong River Basin has evolved. As argued by scholars researching transboundary river basin governance, before the 1990s, China, the upstream state of the Mekong River Basin, had little motivation to cooperate with other Mekong riparian states [26, 28, 33]. However, contrary to arguments by scholars researching hydro-hegemonies [13, 16], cooperation in the Mekong River Basin among the four lower Mekong Countries, first through the Mekong Committee (MC) and later the Mekong River Commission (MRC), was not driven by either the upstream or downstream hegemons, but by hegemons from outside the Mekong region, particularly the US. The communist threat and the resulting Cold War paradoxes drove the Mekong cooperation and the four lower riparian Mekong states formed a downstream hydro-hegemony to counter China's upstream hegemony. Thus, hydropower dam theory was actualized by the lower Mekong cooperation. As a result, China and the four lower Mekong countries squabbled for many years over the management and utilization of the Mekong water resources. China utilized its upstream hegemon status on the Mekong or Lancang River to assert power over the other riparian states along the Mekong. China built many hydropower dams on the mainstream of the Mekong River in Yunnan Province, which caused severe negative social and environmental impacts downstream.

Within the Lower Mekong River Basin Cooperation, the hydro-hegemonic dynamics drive the cooperation. Riparian countries in the Lower Mekong basins cooperate to manage, utilize, and share river resources [14], but they compete in hydropower dam construction as well. For instance, as an upstream hegemon of the Lower Mekong River Basin, Laos continues to build hydropower dams and sells the electricity to Cambodia, Thailand, and Vietnam. At the same time, Thailand finances Laos' Xayaburi Dam, and Vietnam finances Cambodia's LS2 Dam. On the other hand, cooperation in the Lower Mekong River Basin is driven by collusion among riparian states as well, in which the weaker and more powerful states exploit the Mekong River in their ways according to their capacity and resources. Further, powerful and wealthy states assist the poorer and weaker state to exploit the Mekong resources on a larger scale than the weaker states would be able to manage on their own in return for a large share of the proceeds. The consequences of such overexploitation risk the sustainability of the Mekong River Basin.

The cooperation in the Mekong is also driven by external actors, such as the ADB, which has engaged China through the GM's Initiative. Along these lines, hydropower dam projects have been utilized by states and regional hegemons to effect cooperation, conflict, and competition in the region. One of these hydropower projects is the LS2 Dam. Through the GMS, the ADB leverages the hydro-hegemons, such as China and Vietnam to capitalize on the hydropower dam projects in the 3S and the Mekong basins as a strategy to compel resource exploitation, forest clearance, land grants (economic land concessions), and license issuance for agro-plantations, mineral extraction, and infrastructure development.

This chapter has detailed how the US influenced Cambodia and several other riparian states in the Mekong region to come together and first identified the LS2 as a potential hydroelectric development project. US hegemony was carried forward through the continued development of the LS2 project by the ADB/GMS, then the CLV Development Triangle, and the Chinese company. In the end, it was China that successfully competed with Vietnam for a dominant role in the LS2 project and, through this role, China has harvested the bounty from the seeds planted by the US in the 1960s.

These processes transform natural resources and public entities into vehicles for the production and accumulation of wealth for the state, private sector, and powerful political elites in each riparian state. The capital accumulation process binds the state and private interests in the so-called development processes that cut across sectoral boundaries and local communities. Those who criticize these developments are called out for being "anti-development". The impacts of these development patterns extend beyond the rivers, fisheries, and political boundaries into forests, wildlife, spiritual traditions, uplands, and mountains.

Cooperation operates as a factor within power relations, power asymmetries, hydro-hegemony as well as the interests of strong and weak hegemons, which might together be termed "riparian realpolitik". Cambodia, as a weak downstream riparian state in the Mekong, has reoriented itself several times from the 1960s through the present to adjust its interests with hegemonic interests to better its position. This has held over the course of the Kingdom's regime change from monarchy to the killing fields, to totalitarian, socialist, and finally a democratic state. Even now, Cambodia cannot escape from this type of cooperation. Rather, it continues to cooperate with riparian states in the Mekong River Basin to manage, use and share river resources, while at the same time, navigating conflicts, competing for advantage, and colluding with others to exploit the river resources. Thus, cooperation, collusion, and competition enable the riparian states in the Mekong region to extract resources, depending on the country's capacity, economic power, geography, and resource endowment. On the one hand, the smaller, weaker, and poorer countries might exploit resources in one way, while richer and more powerful states exploit these resources in others. On the other hand, cooperation can be used by powerful and rich countries to collude with poorer and weaker states to exploit resources on a larger scale than they could manage in isolation, and, in return, they share the benefits. The consequences of overexploitation of river resources might risk the sustainability of the 3S and the Mekong, thereby imperiling the future of all of these countries.

Through these processes, riparian states politicize relationships, positions, and power relations through cooperation. Within the context of riparian cooperation downstream states act as upstream states, such as Vietnam's efforts to build more dams in the 3S region in central Vietnam, and an upstream state, such as Laos, reinforces its position to build hydropower dams, regardless of opposition from Vietnam and other riparian states. At the same time, Cambodia is climbing the energy ladder, swapping its passive position as a mere complainant for that of a dam builder, with the LS2 Dam set to begin operations soon and rapidly underway for developing the Sambo Dam. These developments are made possible because of China's presence in the region.

Cambodia is vulnerable to regional cooperation. The chapter points to three key issues as ways to counter the hegemony of individual states: the lack of jurisdiction of the MRC on the tributaries, the lack of formalized cooperation between state and non-state actors, and the inability of weak states to counter the strength of hydro-hegemons, not simply because of their weaker power but due as well to their narrower access to information. Building dams is one option for addressing these challenges and the LS2 dam proof that Cambodia can compete for water with other riparians. However, the weaponizing of water in the 3S Basin is about the competition to build dams to retain position and negotiating power in transboundary cooperation, without which Cambodian voices will not be heard.

While China financed the building of the LS2 dam and leveraged the downstream position of Cambodia to compete in the dam building in the Mekong, the Chinese dam has caused social and environmental impacts on local communities living all along the rivers of the basin. As a result, local resistance to the LS2 dam has occurred, as communities seek solutions from the dam builders and government to the problems caused by the development. Lack of experience by the government of Cambodia in addressing the impacts of the dam on local communities and the unaccountability of the Chinese dam builders to social and environmental impacts have only led to an increase in critical local resistance to Chinese dam builders. The local resistance has publicly painted the Chinese in a bad light concerning the LS2 Dam development. If China does not pay attention to improving its performances in overseas investment and development, particularly in Cambodia, local resistance to Chinese investments will increase.

Author details

Mak Sithirith
Department of Natural Resources Management, Faculty of Development Studies, Royal University of Phnom Penh, Cambodia

*Address all correspondence to: maksithirith@gmail.com

References

[1] Daming H, Hsiang-te K. Facilitating regional sustainable development through integrated multi-objective utilizing management of water resources in the Lancang-Mekong River basin'. The Journal of Chinese Geography. 1997; 7: 9-21.

[2] ADB (Asian Development Bank). Strategy and Action Plan for the Greater Mekong Subregion Southern Economic Corridor. Manila: ADB; 2010. Available on https://www.adb.org/sites/default/files/publication/28006/gms-action-plan-south.pdf.

[3] Constable D. Atlas of the 3S Basins. Bangkok: IUCN; 2015. Available on https://portals.iucn.org/library/sites/library/files/documents/2015-064.pdf.

[4] Arias M. E, Piman T, Lauri H. K, Cochrane T. A, and Kummu M. Dams on Mekong Tributaries as Significant Contributors of Hydrological Alternations to the Tonle Sap Floodplain in Cambodia. Hydrology and Earth System Science. 2014; 11: 2177-2209. https://doi.org/10.5194/hess-18-5303-2014.

[5] International Union for Conservation of Nature (IUCN). Profile: The Sesan Basin. Fact Sheet. Bangkok: IUCN, 2015.

[6] Mekong River Commission (MRC). Fisheries. Available at http://www.mrcmekong.org/topics/fisheries/, accessed on 12 November 2019.

[7] World Wildlife Fund (WWF). More than a quarter of a million people say no to Don Sahong dam, 11th September 2014, available at http://wwf.panda.org/?228618/More-than-a-quarter-of-a-million-people-say-no-to-Don-Sahong-dam, accessed on 27 September 2019.

[8] Mekong River Commission (MRC). MRC Master Catalogue. Vientiane: MRC; 2009. Available at http://portal.mrcmekong.org/master-catalogue. Accessed on 15 August 2017.

[9] Piman T, Cochrane T, Arias M. E, Green A. and Dat Nguyen A. Assessment of Flow Changes from Hydropower Development and Operations in Sekong, Sesan, and Srepok Rivers of the Mekong Basin. Journal of Water Resources Planning and Management. 2014; 139: 723-732. DOI: 10.1061/(ASCE)WR.1943-5452.0000286.

[10] Wolf A. Conflict and Cooperation along International Waterways. Water Policy. 1998; 1:251-265. https://doi.org/10.1016/S1366-7017(98)00019-1.

[11] Homer-Dixon T. Environmental Scarcities and Violent Conflict. International Security. 1994; 19:1. https://doi.org/10.2307/2539147.

[12] Gleditsch P. N, urlong K, Hegre H, Lacina B, and Owen T. Conflict Over Shared River: Resource Scarcity or Fuzzy Boundaries. Political Geography. 2006; 25: 361-382. https://doi.org/10.1016/j.polgeo.2006.02.004.

[13] Dinar S. Power Asymmetry and Negotiations in International River Basins. International Negotiation. 2009; 14: 329-360. https://doi.org/10.1163/157180609X432851.

[14] Pearse-Smith S. Water War in the Mekong Basin? Asia Pacific Viewpoint. 2012; 53: 147-162. https://doi.org/10.1111/j.1467-8373.2012.01484.x.

[15] Onishi K. Reassessing Water Security in the Mekong: The Chinese Rapprochement with Southeast Asia. Journal of Natural Resources Policy Research. 2011; 3: 393-412. https://doi.org/10.1080/19390459.2011.607962

[16] Zeitoun M, & Warne J. Hydro-hegemony-a framework for the analysis

of trans-boundary water conflicts. Water Policy. 2006; 8: 435-460. https://doi.org/10.2166/wp.2006.054.

[17] Zeitoun M, and Mirumachi N. Transboundary water interaction 1: reconsidering conflict and cooperation. International Environmental Agreements: Politics, Law, and Economics. 2008; 8:4. https://doi.org/10.1007/s10784-008-9083-5.

[18] Zeitoun M, Mirumachi N, and Warner J. Transboundary water interaction II: Soft power underlying conflict and cooperation. *International Environmental Agreements.* 2011; 11: 159-178. DOI: 10.1007/s10784-010-9134-6.

[19] Pachova N, Nakayama M, and Jansky L. National sovereignty and human security: Changing realities and concepts in international water management. 2008.

[20] Cox R. H, and Schilthuis A. Hegemony and Counter-Hegemony. Encyclopedia of Globalization. London: The Wiley Blackwell; 2012.

[21] Dinar S. *Assessing side-payment and cost-sharing patterns in international water agreements*: The geographic and economic connection. *Political Geography.* 2006; 25: 412-37. https://doi.org/10.1016/j.polgeo.2006.03.007.

[22] Zawahri N. A. Third-party mediation of international river disputes: Lessons from the Indus River. International Negotiation. 2009; 14:281-310. https://doi.org/10.1163/157180609X432833.

[23] Pratt N. Bringing politics back in: examining the link between globalization and democratization, *Review of International Political Economy*. 2004; 11: 331-336. Available online: https://www.jstor.org/pss/4177500. Accessed on 23 July 2017.

[24] Sneddon C, and Fox C. Inland Capture Fisheries and Large River Systems: A Political Economy of Mekong Fisheries. Agrarian Change. 2012; 12: 279-299. DOI: 10.1111/j.1471-0366.2011.00350.x.

[25] Hirsch P. The Shifting Regional Geopolitics of Mekong Dam. Political Geography. 2016; 51:63-74. DOI: 10.1016/j.polgeo.2015.12.004.

[26] Han H. China, an Upstream Hegemon: A Destabilizer for the Governance of the Mekong River? Pacific Focus. 2017; 32:30-55. https://doi.org/10.1111/pafo.12085.

[27] Burgos S. C, and Sophal E. The Hungry Dragon: How China's Resource Quest is Reshaping the World. London & New York: Routledge; 2010.

[28] Liebman A. Trickle-down hegemony? China's 'peaceful rise' and dam building on the Mekong. Contemporary Southeast Asia. 2005; 27: 281-304. https://www.jstor.org/stable/25798737.

[29] Sithirith M. Dams and State Security: Damming the 3S Rivers as a Threat to Cambodian State Security. Asia Pacific Viewpoint. 2016; 57: 60-75. https://doi.org/10.1111/apv.12108.

[30] Hori H. The Mekong: Environment and Develo*pment*. Tokyo: United Nations University Press; 2000.

[31] Ojendal J. Sharing the Good: Mode of Managing Water Resources in the Lower Mekong River Basi*n*. Goteborg: Department of Peace and Development Research, Goteborg University; 2000.

[32] White G. F. The Mekong River Plan. Scientific American. 1963; 208: 49-59.

[33] Jacobs, J. W. The Mekong River Commission: transboundary water resources planning and regional security. The Geographical Journal. 2002; 168: 354-364. DOI: 10.1111/j.0016-7398.2002.00061.x.

[34] Chandler D. A History of Cambodia. Bangkok: Silkworm; 1992.

[35] ADB. The GMS Beyond Borders: Regional Cooperation Strategies and Program 2004-2008. Manila: ADB; 2004. Available on https://think-asia.org/handle/11540/4931.

[36] ADB. The Greater Mekong Subregion Economic Cooperation Program. GMS Assistance Plan (2001-2003). Manila: ADB, 2000. https://www.adb.org/documents/gms-assistance-plan-2001-2003-greater-mekong-subregion-economic-cooperation-program.

[37] Radosevich G, & Olson D. Existing and Emerging Basin Arrangements in Asia: Mekong River Commission Case Study. Third Workshop on River Basin Institution Development June 24, 1999, the World Bank, Washington, DC. http://siteresources.worldbank.org/INTWRD/918599-1112615943168/20431963/MekgongRiverComCaseStudy.pdf. Accessed on 20 November 2018.

[38] MRC. Agreement on the Cooperation for the Sustainable Development of the Mekong River Basin, 5 April 1995. Bangkok: Mekong River Commission Secretariat; 1995.

[39] Mekong River Commission (MRC). Transboundary river basin meeting: 3S basins. Available online: http://www.mrcmekong.org/news-and-events/events/trans-boundary-river-basin-meeting-3s-basins. Accessed on 12 September 2017.

[40] Cambodia-Laos-Vietnam Development Triangle Area (CLV-DTA). (2010). Reporting on reviewing, adjusting, and supplementing the Master Plan for Socio-economic development in Cambodia-Laos-Vietnam Development Triangle Area up to 2020. Available on file:///C:/Users/maksith/Downloads/RevisedMasterplanCLV.PDF, accessed on 13 June 2019.

[41] Grundy-Warr C. B/ordering nature and biophysical geopolitics: A response to Hirsch. Political Geography. 2016; 58: 131-135. https://doi.org/10.1016/j.polgeo.2016.10.006.

[42] The Nation. Water Diplomacy by China Offers Drought Relief. The Nation, Asia News Network; 2016. Available online: www.nationmultimedia.com/news/national/aec/30281969. Accessed on 17 May 2017.

[43] Matthews, N. & Motta, S. Chinese State-Owned Enterprise Investment in Mekong Hydropower: Political and Economic Drivers and Their Implications across the Water, Energy, Food Nexus. Water. 2015; 7: 6269-6284. https://doi.org/10.3390/w7116269.

[44] Phnom Penh Post. Kings of concessions. 25 February 2011a. Available on: http://www.phnompenhpost.com/national/kings-concessions. Accessed on 25 July 2017.

[45] Hensengert O. Regionalism, Identity, and Hydropower Dams: The Chinese-Built Lower Sesan 2. Dam in Cambodia. Journal of Current Chinese Affairs. 2017; *46*: 85-118. https://nbn-resolving.org/urn:nbn:de:gbv:18-4-11097.

[46] Chu T. W. Riparians versus the State in Southeast Asia. Asian Survey. 2017; 57: 1086-1109. DOI:10.1525/as.2017.57.6.1086

[47] Baird I. Non-government Organizations, Villagers, Political Culture, and the Lower Sesan 2 Dam in Northeastern Cambodia. Critical Asian Studies. 2016; 48: 257-277. https://doi.org/10.1080/14672715.2016.1157958.

[48] Mekong Secretariat. *Indicative Basin Plan Report*. Bangkok: Mekong Committee; 1970.

[49] Halcrow and Partners. Sekong, Sesan, and Nam Theun River Basins

Hydropower Development Study: Final Report. Manila: Asian Development Bank; 1999.

[50] ADB. Se Kong-Se San and Nam Theun River Basins Hydropower Study," Final Draft Report. Volume I, November. Manila:ADB; 1998. Available on: Http://s3-us-west-2.amazonaws.com/ucldc-nuxeo-ref-media/8db9ea80-3578-4de5-b2a3-5375831a914d.

[51] Ojendal J, Mathur V, and Sithirith M. Environmental Governance in the Lower Mekong: Hydropower Site Selection Process in the Sesan and Srepok Basins. Stockholm: Stockholm Environment Institute (SEI); 2002. https://www.academia.edu/3718606/Environmental_Governance_in_the_Mekong_Hydropower_Site_Selection_Process_in_the_Se_San_Sre_Pok_River_Basin.

[52] EVN (Electricité du Viet Nam). *Summary Report: Lower Sesan 1/Sesan 5 hydropower project.* Vietnam: EVN; 2010.

[53] Pkeakdey H. China's role in the Cambodian energy sector: Catalyst or antagonist for development? *Southeast Asia Research*. 2015; 23: 405-422, DOI: 10.5367/sear.2015.0272.

[54] Men P, Thun V, Yin S, and Lebel L. Benefit-sharing from Kamchay and Lower Sesan 2 hydropower watersheds in Cambodia. Water resources and rural development. 2014; 4:40-53. https://doi.org/10.1016/j.wrr.2014.10.004.

[55] Key Consultants Cambodia (KCC). Environmental Impact Assessment for Feasibility Study of Lower Sesan 2 Hydropower Project. Phnom Penh: KCC; 2009.

[56] RGC. Law on Cambodian Government Guarantee of Payments to Hydro Power Lower Sesan 2 Dam. Phnom Penh: Royal Government of Cambodia; 2013. Available online: www.internationalrivers.org/sites/default/files/attached-files/draft_law_on_rgc_approval_of_lower_sesan_2_eng-2.pdf. Accessed on 25 September 2017.

[57] Phnom Penh Post. Logging at Dam Site Halted. 12 October 2013, by Khouth Sophak Chakrya, Cheang Sokha, and Daniel Pye, accessed on 15 August 2019, www.phnompenhpost.com/national/logging-dam-sitehalted.

[58] Phnom Penh Post. *Electricity for Vietnam, 1 June 2011b.* available online: http://www.phnompenhpost.com/national/electricity-vietnam, accessed on 15 July 2017.

[59] PECCI and KCC (Power Engineering Consulting Joint-Stock Company No1 and Key Consultants Cambodia). Environmental Impact Assessment for Feasibility Study of Lower Sesan 2 Hydropower Project, Stung Treng Province, Cambodia. Phnom Penh: KCC; 2008.

[60] Work C. Chthonic Sovereigns? 'Neak Ta' in a Cambodian Village. The Asia Pacific Journal of Anthropology. 2019: 20: 79-94. https://doi.org/10.1080/14442213.2018.1553205.

[61] Baran, Eric; Samadee, Saray; Shwu Jiau, Teoh, and Thanh Cong, Tran. Fish And Fisheries in the Sekong, Sesan and Srepok Basins (3s Rivers, Mekong Watershed), with Special Reference to the Sesan River. Project report: Challenge Program on Water & Food Mekong project MK3 "Optimizing the management of a cascade of reservoirs at the catchment level". ICEM: Hanoi, Vietnam; 2013. Available on file:///C:/Users/maksith/Downloads/F_E-2.-Fish-and-fisheries-in-the-Sesan-river-basin.pdf.

[62] SWECO. Environmental Impact Assessment on the Cambodian part of the Sesan River due to Hydropower Development in Vietnam. SWECO Groner (in association with Norwegian Institute for Water

Research, ENVIRO-DEV, and ENS Consult, and Electricité du Viet Nam, December 2006.

[63] Ziv G, Baran E, Nam S, Rodríguez-Iturbe I, & Levin S. A. Trading-off fish biodiversity, food security, and hydropower in the Mekong River Basin. Proceedings of the National Academy of Sciences. 2012: 109:5609-5614. Https://doi.org/10.1073/pnas.1201423109.

[64] Phnom Penh Post. The RGC has granted Hydro Power Lower Sesan 2 Co., Ltd., a joint company created by the cooperation of local companies and companies from Vietnam and a big hydropower company from China, to invest $781.52 million, 24 February 2014.

Section 2

Water Quality Analysis Procedures and Protection Measures

Chapter 4

uMngeni Basin Water Quality Trend Analysis for River Health and Treatability Fitness

Innocent Rangeti and Bloodless (Rimuka) Dzwairo

Abstract

One of the main challenges facing the potable water production industry is deterioration of the quality of raw water. Drinking water that does not meet quality standards is unfit for consumption. Yet, this quality is a function of various factors, key among them being quality of the raw water from which it is processed. This is because costs related to potable water treatment are related to the nature of raw water pollutants and the degree of pollution. Additionally, survival of aquatic species depends on self-purification of the water bodies through attenuation of pollutants, therefore, if this process is not efficient it might result in dwindling of the aquatic life. Hence, this chapter presents spatial and temporal water quality trends along uMngeni Basin, a critical raw water source for KwaZulu-Natal Province, in South Africa. As at 2014 the basin served about 3.8 million people with potable water. Results from this study are discussed in relation to uMngeni River's health status and fitness for production of potable water treatment. Time-series and box plots of 11 water quality variables that were monitored at six stations over a period of eight years (2005 to 2012), were drawn and analysed. The Mann Kendall Trend Test and the Sen's Slope Estimator were employed to test and quantify the magnitude of the quality trends, respectively. Findings showed that raw water (untreated) along uMngeni River was unfit for drinking purposes mainly because of high levels of *Escherichia coli*. However, the observed monthly average dissolved oxygen of 7 mg/L, that was observed on all stations, suggests that the raw water still met acceptable guidelines for freshwater ecosystems. It was noted that algae and turbidity levels peaked during the wet season (November to April), and these values directly relate to chlorine and polymer dosages during potable water treatment.

Keywords: water quality trend, uMngeni, river health, treatability, spatio temporal, water quality, monitoring

1. Introduction

Water is an integral natural component of any ecosystem as well as a social and economic good [1, 2]. However, utilisation of this resource is highly influenced by its quantity and quality. For example, due to limited capacity and the prohibitive cost of treatment chemicals, water treatment utilities and municipalities, especially in developing countries, struggle to provide water for domestic use. This is particularly severe in rapidly growing informal settlements and peri-urban areas. A key factor affecting the production of adequate potable water has been deterioration of the quality of raw water sources [3–5]. This has mainly been attributed to anthropogenic factors [6] such as rapid urbanisation, industrial waste [7], population growth, and climate change among others. The protection of drinking water sources is thus a critical step in ensuring the production of drinking water at a reasonable price. This is also made possible through the participation of various stakeholders with varying knowledge on water science. The research sector has lately been playing a pivotal role in ensuring that water treatment utilities are updated regarding the state of river basins. Hence, failure to incorporate water quality data in water resource management and infrastructure planning challenges the ability of decision makers to respond to the increasing economic and social demands of associated use and consumption of poor-quality water.

The management of water quality is a complex process because of the diversity of polluting parameters as well as their interaction and resultant characteristics within a polluted environment [2]. An important aspect of water quality assessments for drinking water source protection is identifying changes in water quality trends and specific pollution sources, which are an integral process during appraisals of decisions related to environmental protection [8, 9]. To assist with these appraisals, long-term water quality data are employed as they are useful for detecting the spatial and temporal changes in the state of aquatic environments, when interpreting effectiveness of measures aimed at [10]. Water quality monitoring can thus assist to reveal the history of a water body and further give a reflection of the current status. Water quality trends analysis present a simple method of determining the long-term fitness of a catchment for a given use. The chosen technique is expected to assist with identifying how the concentration of a given water parameter changed over a period of time. With this method, one can determine the long-term fitness of a water body for specific uses such as potable production and protection of aquatic biodiversity. The success of a trend analysis depends on the main principles (1) acquisition of water quality data from a properly-designed monitoring program, (2) application of appropriate statistical methods for trend detection, and (3) a good understanding of relevant water quality parameter correlations [11].

It is known that rivers are often the ultimate sinks of effluent and other forms of pollution [12, 13]. Various processes, for example, soil erosion, disposal of waste from urban-industrial and domestic activities including fertiliser run off from agricultural activities, cause the deterioration of water quality [2]. Specifically, polluted water is a threat to flora and fauna and is unsuitable for any human need [14]. Thus, consumption of water that is polluted results in ill health which, in extreme cases, can cause death [4]. Pathogens present in drinking water including many bacterial, viral and protozoan agents have been reported in various research articles to cause illness and death to various degrees and in different year timelines [15–17]. Some specific health and water-related problems have statistically was grouped by Khan, Tareen [18] as follows: hepatitis A (32%–38%), gastroenteritis (40%–50%), hepatitis B (16%–19%), hepatitis C (6-7%), dysentery (28–35%), and diarrhea (47%–59%).

Each water quality variable (parameter) also has an effect, either beneficial or detrimental on aquatic organisms. The effect depends on whether they act synergistically or antagonistically [12]. To aquatic organisms, this effect is influenced by the tolerance limit of that organism. However, the fundamental principle of pollution prevention is to stop or minimise anthropogenic activities, which result in the contamination of water systems. In determining pollution prevention strategies, it is important to firstly identify the key pollutants of concern [19]. South Africa is not exceptional to the current worldwide deterioration of the quality of river waters. It is perceived as one of the major impediments to South Africa's capability to provide sufficient water of appropriate quality to meet its current needs and to ensure sustainable water provision for the future [20]. The quality of water directly affects its use, such that poor quality water is less available for various uses. An understanding of the temporal and spatial changes in the water quality of rivers is critical for water resource protection and management.

During the production of potable water, several chemicals, for example, disinfectants, coagulants and oxidants, are added at various stages of the process [21], the objective being to produce water that is biologically, chemically and aesthetically acceptable. The cost of chemicals is a major component of operating a potable water production plant [22, 23] although the types and quantities vary depending on a range of factors such as incoming raw water quality and the treatment technology [21]. Because water treatment costs determine the pricing of potable water, it means that ultimately cost presents a barrier regarding access to safe and adequate water for drinking and sanitation purposes, especially for poor communities [5, 24]. Additionally, the measured water quality variables, for example, turbidity and pH, have either a direct or indirect effect on the dosage and performance of the treatment chemicals. One condition is when there is high pH water, which will require a comparably lower coagulant dosage since coagulant optimisation is at high pH values. On the other hand, an elevated turbidity level increases coagulant dosage. Coagulant is a chemical that conglomerates dissolved pollutants into flocs that can then be removed using filtration and sedimentation [25]. Thus, the cost of potable water treatment due to diminished water quality represents an important component of the societal costs of water pollution [24, 26]. Therefore, finding timely, cost-effective, and appropriate solutions that address water pollution and its resulting challenges is critical.

Hence, the main objective of this chapter was to highlight water quality trends in uMngeni Basin for the period 2005 to 2012. The focus was on (1) how water quality variability impacted chemical dosage during potable water production at Durban Heights and Wiggins Water Treatment Plant and (2) determining how the quality variability affected the overall freshwater ecosystem health within the basin, using specific parameters. The following sections describe uMngeni Basin, provide the methods employed during analysis as well as results and conclusions drawn.

2. Study area

uMngeni River (232 km) and its basin (4432 km^2) are the primary source of raw water that is then treated to potable use for a population of almost 6.7 million people [27] within and around Durban Metro as well as Pietermaritzburg City, in KwaZulu-Natal (KZN) Province, South Africa (**Figure 1**) uMgeni Water, a State-Owned Entity, is the biggest 'user' of raw water in the uMngeni Basin as it produces 472 million cubic meters of potable water per year for commercial, industrial and domestic use along the basin and its adjacent coastal areas [27]. The catchment receives an annual precipitation of 410–1450 mm, mean annual runoff which varies from 72 mm to 680 mm and a mean annual evaporation of 1360–2040 mm (CSIR., 2011).

uMngeni Basin supports a diverse range of livelihood activities which contribute approximately 16% to the gross national product of South Africa [28]. Although the province covers a small portion of the country's land area, the province contributes almost 30% to the national agricultural output and, therefore, significantly bridges food security in South Africa. A consequence of this development has been high levels of pollutants entering the water resources system and getting flushed out to sea.

The basin's main river rises close to the uMgeni Vlei, near Fort Nottingham at an altitude of 2071 m (**Figure 2**). The land surrounding the headwaters is mainly used for forestry plantation and agriculture [29]. Most of the farmers practice dairy, piggery and maize production [29]. As a result of many economic activities, damage to the

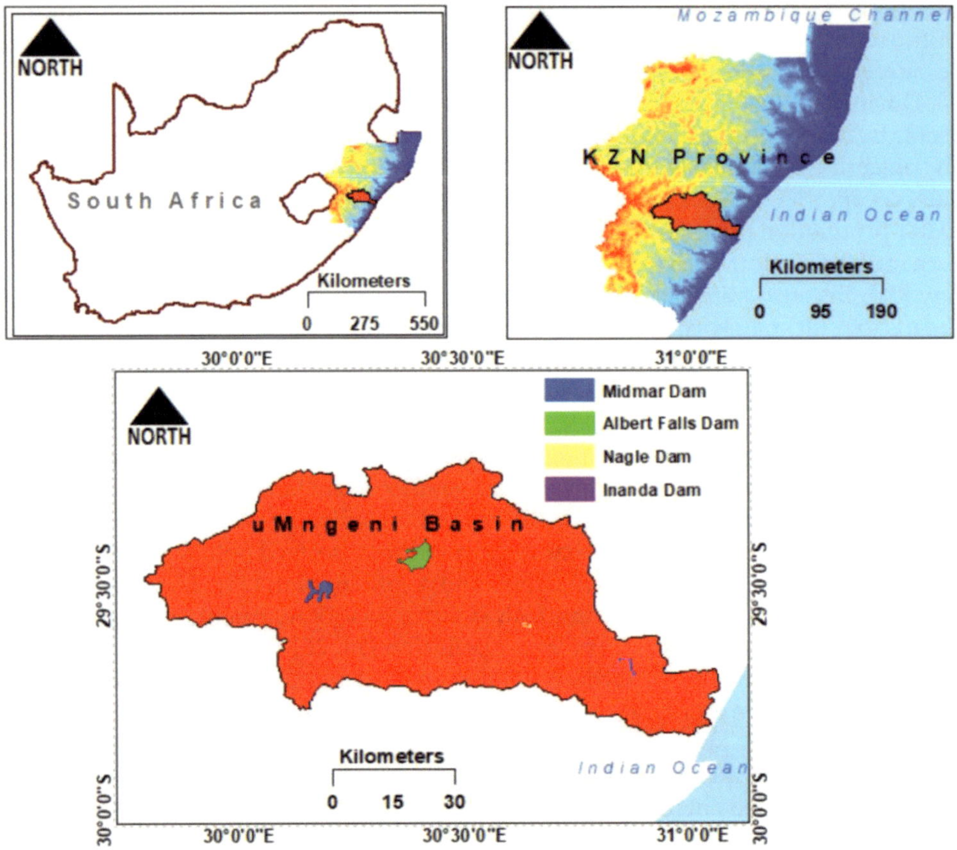

Figure 1.
South Africa, KwaZulu-Natal Province and uMngeni Basin plus four dams.

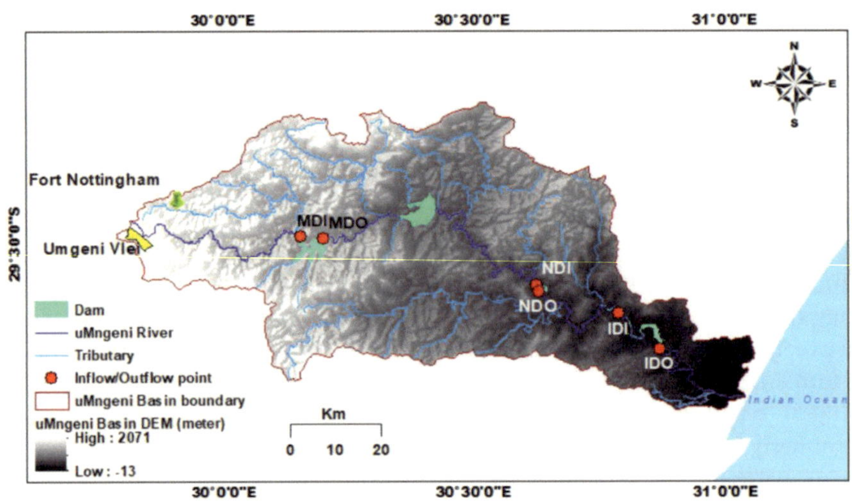

Figure 2.
uMngeni River and dams along its course, showing dam inflows and outflows.

headwater wetlands has been reported [30]. The removal of riparian vegetation and invasion by exotic flora has also contributed to erosion of the riverbank throughout the upper uMngeni River. Four major dams namely; Midmar, Albert Falls, Nagle and

Inanda are located along the river course as it flows towards the Indian Ocean. Except for Albert Falls, raw water is abstracted from the other three dams by Umgeni Water Amanzi, for the purpose of producing potable water. Because Umgeni Water Amanzi's primary business depends on the quality of raw water, the treatment utility works in conjunction with the Department of Water and Sanitation (DWS) to monitor and manage the dams. The monitoring program has generating a vast amount of data that could be used to improve management of the basin, overall. This chapter is based on the analysis of historical data, which was obtained from uMgeni Water (2013).

Station 1: Midmar Dam Inflow (MDI): The station is located at the mouth of Midmar Dam. It measures the quality of inflow that helps to reflect the polluting activities upstream the river. Literature has cited agricultural activities runoffs and sewage bursts from low-cost settlements such as Mpophomeni, as significant polluting activities in the upstream of this dam [31]. Intensive livestock production and its related activities like cattle grazing, poorly managed dairies and piggeries, are the dominant sources of faecal pollution in Midmar and Albert falls dam catchments.

Station 2: Midmar Dam Outflow (MDO): The station is located where the dam re-joins river. It measures the quality of water before being discharged into uMngeni River. This helps reflect the natural purification efficiency of the dam. Since treatment plants generally abstract water of the best quality, this station should indicate the quality expected at Midmar Water treatment plant.

Station 3: Nagle Dam Inflow (NDI): The station is located downstream of Albert Falls Dam, which acts as reservoir for the former. Since it is located at the mouth of Nagle Dam, it measures the inflow from the river and thus helps reflect the pollution activities and river health status between Albert Falls and Nagle dams.

Station 4: Nagle Dam Outflow (NDO): The station is located where Nagle Dam re-joins uMngeni River. The quality of water at this station is expected to give an overview of the raw water being abstracted for treatment at Durban Heights.

Station 5: Inanda Dam Inflow (IDI): The station is located at the mouth of Inanda. It basically reflects the pollution activities from Nagle Dam to Inanda Dam. It gives information on the pollution activities arising due to activities by the rural community in the Valley of Thousand hills. Inanda Dam has been reported to experience some periods of elevated algal counts.

Station 6: Inanda Dam Outflow (IDO): The station is located at the outflow of Inanda into uMngeni River. When compared with its subsequent inflow station, it helps reflect the capacity of the dam to naturally purify water due to retention. Since treatment plants generally abstract from the best quality point, this station is expected to show the treatability of water at Wiggins treatment plant.

Catchment water quality monitoring involves the analysis of various parameters such as biological pathogens (e.g. *E. coli*), physical parameters (e.g. conductivity), metals, organic pollutants, turbidity, total suspended solids and pH [32]. These parameters should fall within specified guideline values for the raw water to be deemed safe for use. Studies have reported that statistic analysis of water quality data, is mainly affected by the nature of the data-sets, which are often non-monotonic trends, non-linearity, non-normally distributed, exhibit seasonal variations and have time spacing that is uneven [9, 33, 34]. This approach informed on the methods and materials for this study.

3. Material and methods

Eight-year data-sets (2005 to 2012) for the six stations as indicated in **Figure 2**, were acquired from the then uMgeni Water (2013) and analysed. Eleven (11) parameters namely alkalinity, pH, electrical conductivity (EC), nitrate, phosphorus, temperature, turbidity, total algae count, *E. coli*, dissolved oxygen and ammonia were finally considered for spatial and temporal trend analysis. These parameters were selected for trend analysis in this study based on the following three criteria; (1) literature citing the parameter as significant to aquatic biodiversity health, (2) the availability of data and (3) the parameter being of interest mainly due to pollution activities in the study area. In addition to raw water quality, potable water quality data-sets for Durban Heights and Wiggins treatment plant for the period 2005 to 2012 were also obtained for comparative analysis. Both graphical (time-series and box-plot) and statistical SK test methods were employed.

3.1 Time-Series Plot

A time series is a set of observations obtained by measuring a single variable regularly over a given period. The main characteristics of a time series analysis are that the data are not independently sampled, their dispersion varies in time, and they are often governed by a trend and cyclic components. In this study, time-series line plots were produced for all parameters in order to depict the variable patterns. Even though they graphically provided an easy and quick method of assessing the pollution patterns, their inability to display the crucial data characteristics, specifically mean, mode and median, was a major drawback. While considering the importance of these characteristics for describing water quality of the basin, box-plots were also done to augment the graphical trending [35]. Since effective trend analysis requires a fairly long sequence of data for a given sampling site, it was considered to use a five-year monthly data-set at a minimum, for monotonic trend analysis. However, for a step-trend, at least two years monthly data before and after treatment would be required [36]. These time frames are only guidelines as the longer the periods of record, the greater would be the sensitivity to detect smaller changes, based on the minimum guideline for monotonic and step.

3.2 Box-plot

One of the main advantages of a box-plot, also known as box and whisker, is its ability to summarise the distribution of data-sets in a way that allows for spatial comparison. This technique gives a visual summary of; (1) the centre of the data (the median = the centre line of the box), (2) the variation or spread (inter-quartile range = the box height), (3) the skewness (quartile skew = the relative size of box halves) and (4) presence or absence of unusual values 'outliers' [35, 37]. In this study the box plot was used to visually capture spatial variation among the study area stations. By allowing a comparison of data without making any statistical assumption, a box-plot presented a greater advantage compared to other observational techniques. However, the main limitation of observational methods was inability to quantify the magnitude of the observed trend [35, 37]. While considering the importance of determining the magnitude of trend especially when forecasting, statistical methods were also employed in this study.

When a study depends on secondary data, it is reasonable to consider a statistical method that accommodates outliers, missing values and censored-values which are common in water quality data-sets. The Seasonal Kendall test (SK test), which is a special case of the Mann Kendall Trend Test, is such a non-parametric test that can handle non-normal distributed data even with outliers, missing values or censored-values [35]. It is used to test for a monotonic trend of a parameter of interest when the data collected over time are expected to change in the same direction (up or down) for one or more seasons, e.g., months. The following assumptions underlie the SK test: (1) When no trend is present the observations are not serially correlated over time, (2) the observations obtained over time are representative of the true conditions at sampling times, (3) the sample collection, handling, and measurement methods provide unbiased and representative observations of the underlying populations over time, (4) any monotonic trends present are all in the same direction (up or down). If the trend is up in some seasons and down in other seasons, the SK test will be misleading, (5) the standard normal distribution may be used to evaluate if the computed SK test statistic indicates the existence of a monotonic trend over time [33].

A positive SK value indicates an increasing trend whilst a negative value shows a decrease. A trend is considered statistically significant if the calculated P - value is less than 0.05 [36]. In this study, for situations where a significant trend was obtained, the Sen's Slope test was then employed to determine the magnitude of the trend. More detailed descriptions of the SK are found in Hirsch, Slack [38], Darken [39] and Helsel and Hirsch [35]. Both the SK and Sen's slope were calculated using XLSTAT 2014 software. The software runs in Excel and does not require prior programming knowledge, making it simple and user-friendly.

4. Results and discussion

4.1 Turbidity

The time series graph depicted in **Figure 3** shows that turbidity exhibited seasonal variation with regular peaks in the wet season (October to March). The most plausible explanation for the peaks is the increased soil and other particulates runoff due to rain. The mean and median values (**Table 1**) for all stations studied shows that turbidity was above the South Africa expected limit for drinking water, which is 0–1 NTU [40]. The SK test presented in **Table 1** shows a decreasing turbidity trend from Nagle Dam inflow station to Inanda Dam outflow. The most plausible explanations for the improvement of quality especially at Inanda Outflow station could be the retention effect which could have occurred as the flow decreased. Midmar Inflow (306 NTU) recorded the highest turbidity level, which could be attributed to intensive farming activities upstream this station.

Chlorination of highly turbid waters tends to increase the level of trihalomethane (THM) which has been reported as carcinogenic [41, 42]. Excessive turbidity in drinking water is aesthetically unappealing and is of health concern as the suspended particles can provide food and shelter for pathogens [43]. If not removed, turbidity can promote regrowth of pathogens in the distribution system and this could lead to waterborne disease outbreaks. The South African guideline requires

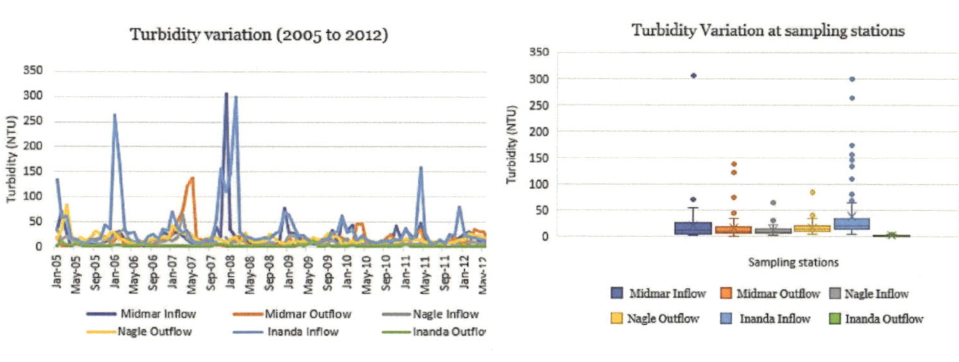

Figure 3.
Turbidity time series and box -plot.

Sample	Minimum	Maximum	Median	Mean	S-K (P)	Sen's slope	Decision
Midmar inflow	2.850	306.000	12.900	20.361	0.385	0.028	No trend
Midmar outflow	0.360	139.000	11.500	16.577	0.211	0.059	No trend
Nagle inflow	3.250	72.740	9.405	12.160	<0.0001	−0.078	Decrease
Nagle outflow	4.910	84.000	14.650	17.286	0.0495	−0.052	Decrease
Inanda inflow	5.635	299.700	21.563	38.729	0.0351	−0.075	Decrease
Inanda outflow	1.040	12.400	2.060	2.683	< 0.0001	−0.012	Decrease

Table 1.
Descriptive statistics for turbidity.

that turbidity level for drinking water be below 5 NTU [21, 44]. Kuutondokwa [45] reported of viable coliform counts in water with turbidity higher than 3 NTU, even in the presence of free residual chlorine. Dearmont, McCarl [26] reported that a 1% increase in turbidity tends to increase chemical costs by one fourth of a percent. Pernitsky and Edzwald [46] further reported that coagulant doses were generally higher when raw water turbidity was high, although the relationship was not linear. To aquatic life, higher than normal turbidity decreases the percentage of light transmitted, which results in decreased photosynthesis [45]. Additionally, excessive turbidity tends to demand an increase in the dosage of disinfectant in order to cater for envisaged pathogens that could be harbouring on the surfaces of the particles. Further, treating water with elevated levels of turbidity naturally results in excess sludge [45], which is undesirable for the environment.

On the other hand, a relatively low turbidity level observed at Inanda Dam Outflow station was desirable for the Wiggins Potable Water Treatment Plant since it meant using minimal quantities of chlorine and polymer. Dennison and Lyne [23], however, while studying factors causing high treatment cost at DV Harris Water Treatment Plant, observed that in contrary, low turbidity levels at Midmar Dam resulted in high treatment cost. The cost was attributed to use of bentonite, a coagulation enhancer that tends to facilitate the coagulation process for rather difficult to 'clean' (less turbid) raw water. The cost of bentonite tends to counteract the reduction of polymer dosage, making it rather expensive to treat water of with turbidity lower than optimal for efficient treatment to potable water quality.

4.2 Algae

For quantifying algae, only dam outflow station data-sets were used. Results for the three stations show that algae is a major problem along uMngeni Basin. As expected, algae exhibited peaks during the wet seasons [47] (**Figure 4**). The mean range of 1900 to 7055 cell/mL indicates that algae were dominant along the basin (**Table 2**). By comparing the mean observations of the three stations it is noted that Nagle Outflow recorded the highest mean concentration among all the three stations.

However, a comparison of the median values (**Table 2**) clearly shows that Inanda is the most deteriorated among the three stations. The latter results conform to earlier studies [29, 48]. During potable water treatment algae is known to have a direct relationship to polymer and chlorine dosage. The high level observed using water from Nagle Outflow suggests that more polymer and chlorine is required to treat water from Nagle compared to that from Inanda Outflow.

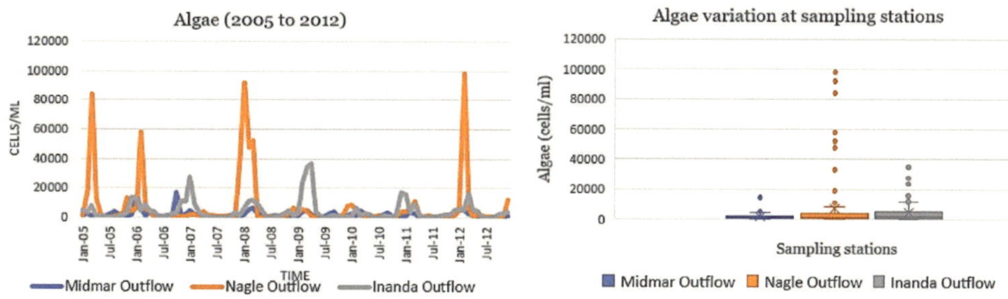

Figure 4.
Algae time series and the box -plot.

Sample station	Minimum	Maximum	Median	Mean	SK (p value)	Sen's	Decision
Midmar outflow	159.500	16760.500	1210.000	1900.600	0.125	−4.64	No trend
Nagle outflow	124.667	98224.400	1155.585	7055.885	0.497	−2.968	No trend
Inanda outflow	105.750	36352.330	1998.750	4499.085	0.748	−1.766	No trend

Table 2.
Descriptive statistics for Algae.

4.3 pH

The time-series plot (**Figure 5**) illustrates that pH variation is not a major problem along uMngeni Basin. Except for Inanda Inflow, the results for the SK test (**Table 3**) reveal no significant temporal variation of pH among the six stations. The observed minimum to maximum ranges were within the general national and international limits for no human risk (6.0 to 9.0 pH) and the survival of fresh water biodiversity [40, 49]. Growth and reproduction of freshwater aquatic species such as fish are found to be ideal within a pH range of 6.5 to 8.5 while pH below 4 or above 10 tends to be lethal to most aquatic animals [21, 50]. The pH values observed would not ideally promote toxicity of dissolved metal ions and protonated species, resulting in aesthetically pleasing water. Additionally, consumption of the water would not cause significant adverse health effects.

The pH of an aquatic system determines the concentration, accumulation and bioavailability of various species, which could be advantageous if required but a disadvantage if this process results in release of pollutants [51]. For example, a

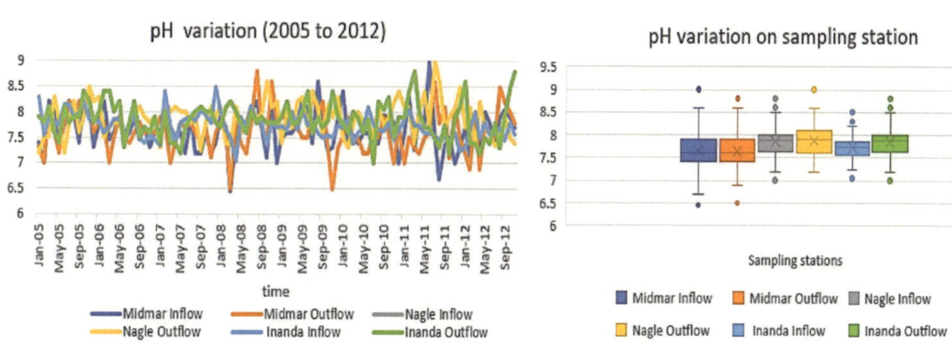

Figure 5.
pH time series and the box-plot.

Sample	Minimum	Maximum	Median	Mean	S-K (P)	Sen's slope	Decision
Midmar inflow	6.450	9.000	7.600	7.646	0.745	0	No trend
Midmar outflow	6.500	8.800	7.600	7.641	0.491	0	No trend
Nagle inflow	7.000	8.800	7.875	7.857	0.77	0	No trend
Nagle outflow	7.200	9.000	7.900	7.880	0.77	0	No trend
Inanda inflow	7.050	8.500	7.742	7.732	0.002	−0.003	Decrease
Inanda outflow	7.000	8.800	7.875	7.857	0.77	0	No trend

Table 3.
Descriptive statistics for pH.

decrease in pH increases the corrosivity of water, and increasing the pH increases the tendency to precipitate mineral scales such as calcium carbonate ($CaCO_3$). Another example relates to ammonia, which is more toxic in alkaline water than in acidic because free (NH_3) at high pH values (pH >8.5) is more toxic to aquatic biota than when in its oxidised form, NH_4^+. Additionally, reduction of pH enhances solubility and speciation of some metals, which elevates their toxicity [52]. On the other hand, alkaline conditions promote precipitation of some cations when they complex with other dissolved ligands [53]. Thus, overall, mobilisation of the species might be beneficial or undesirable, depending on the receiving environment and the concentrations resulting [51].

4.4 Total alkalinity

Alkalinity ranged from a minimum value of 14.6 mg/L at Inanda Inflow to a maximum value of 83.5 mg/L at Inanda Outflow (**Figure 6**) This indicates that uMngeni Basin still had the ability to resist pH change to some degree. Regarding river health, the observed high mean range (27.974 mg/L - 62.842 mg/L) (**Table 4**) suggests that uMngeni River still had the potential to resist pH, resulting in protection of the aquatic animals. This is in agreement with Dallas and Day [54], who reported that South Africa's rivers are naturally alkaline. The SK test results presented in (**Table 4**) show that alkalinity was generally stable during the period 2005 to 2012. A comparison of the mean and median values illustrates that both the inflow and outflow stations showed a moderate spatial increase towards downstream of the uMngeni River. Inanda Inflow station recorded a monthly mean value of 55.840 mg/L which was approximately twice that of Midmar Inflow (28.11 mg/L) (**Table 4**).

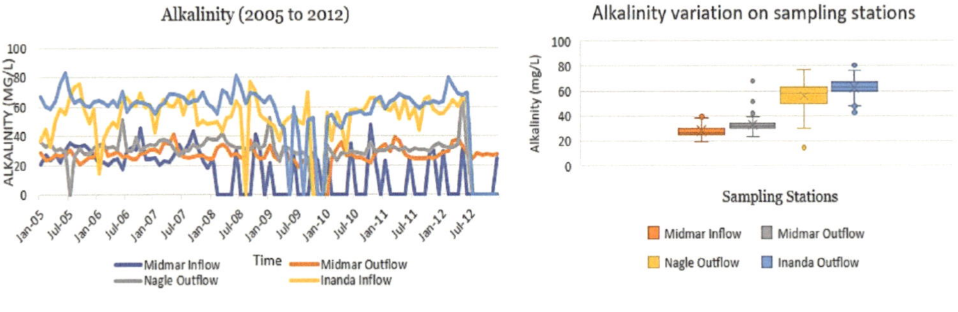

Figure 6.
Total alkalinity time series and the box-plot.

Sample	Minimum	Maximum	Median	Mean	S-K	Sen's slope	Decision
Midmar inflow	17.530	47.500	26.850	28.110	0.492	0.026	No trend
Midmar outflow	19.160	41.400	26.650	27.974	1	0.004	No
Nagle outflow	23.280	67.900	31.815	32.657	0.461	−0.018	No
Inanda inflow	14.600	77.270	57.695	55.840	0.363	0.077	No
Inanda outflow	42.970	83.500	62.995	62.842	0.633	−0.015	No

Table 4.
Descriptive statistics for Total alkalinity.

Regarding potable water production, the observed high alkalinity level at Inanda Dam stations reflects its ability to resist pH change as a result of coagulation. This could explain the low lime dosage used at Wiggins Treatment Plant compared to Durban Heights Treatment Plant.

4.5 Temperature

As expected with South African seasons [47], temperature showed seasonal variation with fluctuations (7.2–12°C) in winter (May – August) and high temperatures (24.7–29.9°C) in summer (September to March) (**Figure 7**) The small variation at the stations at each given time of the year could be explained by the differences in sampling time. The SK and Sen's slope results in **Table 5** show that water temperature decreased slightly towards downstream along uMngeni Basin except at Midmar Outflow.

High temperatures observed in summer coincided with algae bloom. Graham [48] used a data range from 1990 to 1999 to explain that algae bloom was a significant driver for potable water treatment costs along uMngeni Basin. Low temperatures have been reported to result in the formation of small flocs, which are vulnerable to disintegration due to fluid shear force [55]. On the other hand, Veenstra and Schnoor [56] explained that high chemical dosages for potable water treatment during summer were meant to compensate for losses caused by the sun's radiation, making it more expensive to treat water in summer. Even then, since very little can be done operationally to remedy temperature variation, a treatment plant should have strategies in place to counter the effect of low temperature in winter, especially for coagulation and flocculation processes.

Additionally, temperature tends to influence the metabolic rates of aquatic organisms, which influence the quality of the raw water. In warm waters respiration rate tends to increase, leading to increased oxygen consumption and decomposition of organic matter. The concentration of oxygen in water is also affected by temperature,

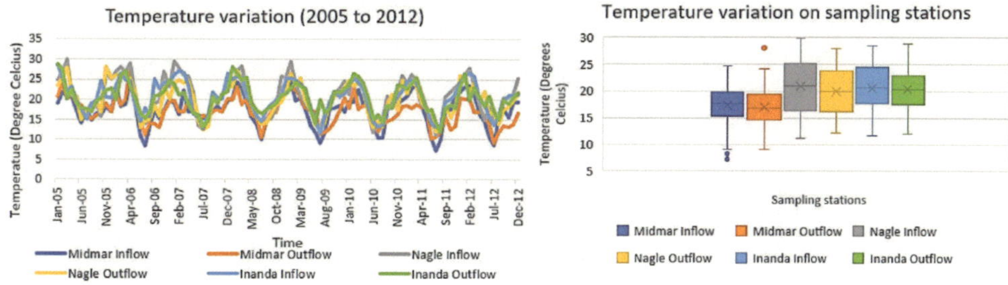

Figure 7.
Temperature time series and the box-plot.

Sample	Minimum	Maximum	Median	Mean	S-K	Sen's slope	Decision
Midmar inflow	7.200	24.700	17.700	17.397	0.83	−0.003	No
Midmar outflow	9.100	28.100	16.900	17.029	< 0.0001	−0.038	Decrease
Nagle inflow	11.200	29.900	21.000	20.908	0.0455	−0.029	Decrease
Nagle outflow	12.100	28.000	19.975	19.964	0.001	−0.031	Decrease
Inanda inflow	11.633	28.450	20.750	20.587	0.002997	−0.026	Decrease
Inanda outflow	12.000	28.800	20.400	20.419	0.0379	−0.013	Decrease

Table 5.
Descriptive statistics for Temperature.

with increasing temperature encouraging escape of the gas from water. The rate of biodegradation of organic compounds increases with increased temperature and this further adds to the reduction of dissolved oxygen as well as nutrient accumulation [57]. Metals and compounds behave differently in low and high temperature environments. For example, $CaCO_3$ precipitates out in hot water to form scaling, which can cause clogging in hot water pipes [58]. Therefore, overall, temperature affects the quality of raw water that feeds into a potable water treatment plant.

4.6 Dissolved oxygen

Dissolved oxygen levels varied between 4.2 mg/L and 11.80 mg/L as depicted in **Figure 8**. Except for Inanda Outflow, the SK test results depicted in the table show that dissolved oxygen remained fairly stable among the stations studied between 2005 and 2012. The high mean (7.926 mg/L - 9.385 mg/L) and median (8.100 mg/L - 9.500 mg/L) (**Table 6**) range shows that dissolved oxygen along uMngeni Basin was within the 80–100% assuming 8.0 mg/L as saturation level [59]. Dissolved oxygen concentrations in unpolluted water normally range between 8 and 10 mg/L [60].

The minimum values observed at Midmar Dam, which are below 5 mg/L (**Table 6**) is worrisome when considering fitness for the survival of aquatic biodiversity. Low dissolved oxygen concentrations are known to adversely affect

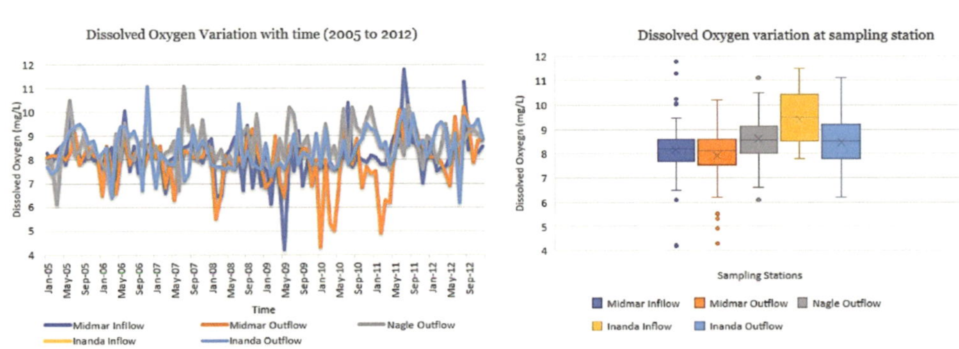

Figure 8.
Dissolved oxygen time series and the box-plot.

Sample	Minimum	Maximum	Median	Mean	S-K	Sen's slope	Decision
Midmar inflow	4.200	11.800	8.175	8.153	0.6	0.002	No
Midmar outflow	4.300	10.200	8.100	7.926	0.7734	0	No
Nagle outflow	6.100	11.100	8.550	8.608	0.0658	0.007	No
Inanda inflow	7.800	11.500	9.500	9.385		−0.063	
Inanda outflow	6.200	11.100	8.500	8.457	0.036	0.006	Increase

Table 6.
Descriptive statistics for dissolved oxygen.

the performance and survival of aerobic organisms while levels below 2 mg/L may be fatal to fish. Regarding treatment, the causal relationship between dissolved oxygen and chemicals used during treatment is not well understood. Some reports have, however, highlighted that oxygen tends to attach to the floc particles making them lighter and relatively impossible to settle for easy flocculation.

4.7 Nitrate

The time-series plot (**Figure 9**) shows that nitrate exhibited an irregular pattern at the six stations studied. This could be attributed to many factors key among them being the irregular discharge of poor-quality sewage effluent along the basin. High concentrations observed in winter tended to be a response of reduced flows, which affected the dilution capacity. Peaks observed during the wet season can be explained by the increased surface run-off, which would encourage movement of fertilisers, animal feedlots effluent and increased sewage effluent discharge. Even though the results indicated that nitrate levels (**Table 7**) were within the regulatory limit for potable use (<11 mg/L), it is important to note that the parameter is the cause of eutrophication especially if levels exceed the recommended limits for no risk, i.e., 0 to 0.5 mg/L as N [44].

Under eutrophic conditions, dissolved oxygen greatly increases during the day, but is greatly reduced after dark by the respiring algae and microorganisms that feed on the elevated mass of dead algae. Oxygen is required by all aerobically respiring plants and animals and it is replenished during daylight by photosynthesising plants and algae. When dissolved oxygen levels decline to hypoxic (inadequate) levels, fish and other aquatic organisms that require oxygen suffocate. These challenges could cause odour and taste problems as well as death of aquatic animals. Additionally, algal blooms limit sunlight penetration to bottom-dwelling organisms and cause wide swings in the amount of dissolved oxygen in the water.

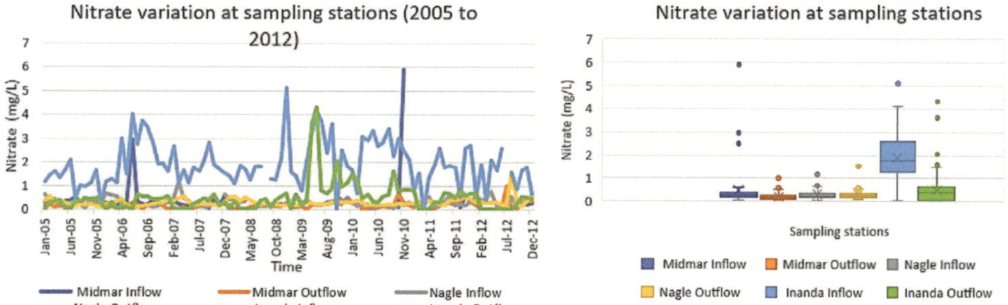

Figure 9.
Nitrate time series and the box-plot.

Sample	Minimum	Maximum	Median	Mean	Sen's slope	Decision
Midmar inflow	0.038	5.910	0.270	0.467	−0.001	No trend
Midmar outflow	0.038	1.000	0.165	0.197	2.45E-04	No trend
Nagle inflow	0.038	1.180	0.240	0.299	0	No trend
Nagle outflow	0.090	1.520	0.260	0.294	8.54E-04	Trend
Inanda inflow	0.038	5.120	1.780	1.910	0.003	No trend
Inanda outflow	0.038	4.330	0.385	0.492	0.002	Trend

Table 7.
Descriptive statistics for nitrate.

Nitrate (NO_3^-) as depicted in (**Figure 9**) could account for algae blooms in the wet season, which would ultimately cause increased chlorine and coagulant dosages during potable water treatment. However, the nitrogen levels along uMngeni Basin are not considered to pose a problem to communities when the receiving water bodies are used for domestic and recreational purposes. It is important to maintain the levels low as studies have reported that (NO_3^-) concentration above the permissible limits could be lethal to infants by causing the "blue baby" syndrome in bottle-fed babies [15, 61, 62].

4.8 Total phosphate

Just like nitrate, total phosphate showed irregular variation with peaks throughout the year (**Figure 10**). This suggests that the two variables could be emanating from the same sources. The most plausible sources could be fertilisers and sewage effluent pollutants from anthropogenic activities along the river. Furthermore, while considering that the soils of uMngeni Catchment are generally phosphorus deficient as highlighted by Furness and Richard [63], it is reasonable to suggest that human activities could be responsible for the peaks. The SK test reveals that there were no significant trends in total phosphate values at any of the stations during the study period (**Table 8**). Descriptive statistic results (**Table 8**) illustrate that Inanda Inflow station had comparably twice (130.078 μg/L) the total phosphate concentration as that of Nagle Inflow station (57.798 μg/L) which is upstream station. Darvill Wastewater Treatment Works (WTW) treats domestic and industrial effluent from Pietermaritzburg, contributing a significant nutrient load (total phosphorus is 15% and soluble phosphorus 50%) and high oxygen demand to Inanda Dam.

Regarding potable water treatment, the effect of total phosphate is indicative of the resultant algae bloom. Thus, in order to reduce the cost associated with the production of potable water along uMngeni River catchment, attention needs to be focused on the reduction of the nutrient load into the river system.

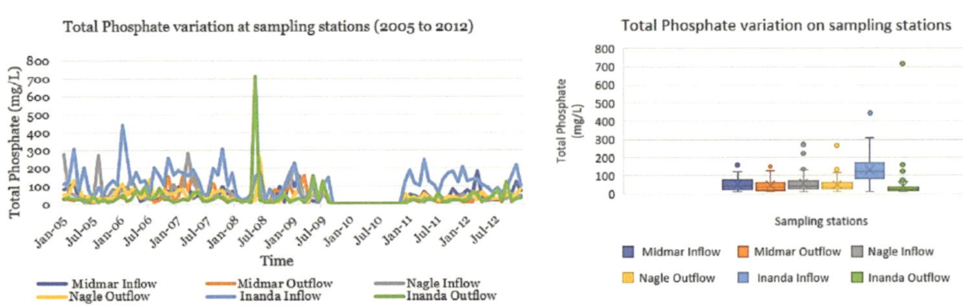

Figure 10.
Total Phosphate time series and the box–plot.

Sample	Minimum	Maximum	Median	Mean	S-K	Sen's slope	Decision
Midmar inflow	11.250	180.977	42.950	53.238	1	−0.065	No trend
Midmar outflow	11.250	161.000	35.200	46.245	0.913	−0.08	No trend
Nagle inflow	11.250	280.550	41.967	57.798	0.116	−0.389	No trend
Nagle outflow	11.250	265.400	39.700	49.922	0.074	−0.241	No trend
Inanda inflow	13.250	445.500	120.500	130.078	0.704	−0.234	No trend
Inanda outflow	11.250	714.300	26.100	42.012	0.347	−0.002	No trend

Table 8.
Descriptive statistics for total phosphate.

4.9 *E. coli*

The time-series plots in **Figure 11** depict seasonal fluctuation and peaks of *E. coli* levels. The peaks that are pronounced in the wet season could be as a result of increased surface runoff from burst sewer pipes and slurry from intensive livestock farming operations that flow into the river course [64, 65]. Storm wash-off containing accumulated human, animal and domestic waste material and overflowing pit-latrines also cause contamination during high flows [64, 65]. The mean range of 97 to 1319 CFU/100 mL (**Table 9**) shows that water quality along uMngeni Basin is not fit for direct drinking purposes without disinfecting. South Africa's guidelines stipulate that there should be zero *E. coli* in 100 mL of the test water [40]. The observed high range could be due to many factors chief among them being the discharge of poor-quality sewage effluent.

With regard to human health, the observed high *E. coli* levels recorded at Inanda Inflow are worrisome when considering that there are nearby rural communities of The Valley of a Thousand Hills, which might use the river water for domestic purposes. Serious faecal contamination problems have been reported, for example, in Pietermaritzburg and Durban, as well as in the settlements in The Valley of a Thousand Hills and Vulindlela (Henley), where Karar [66] reported that annual deaths were associated with water borne diseases. The major sources of pollution were noted to be leakage and blocked sewers in formally serviced residential areas (e.g. Pietermaritzburg and Durban) and illegal disposal of waste into stormwater drains as well as general drainage.

Regarding potable water treatment costing, the elevated levels of *E. coli* observed towards downstream at Inanda added pressure to the Wiggins Potable Water

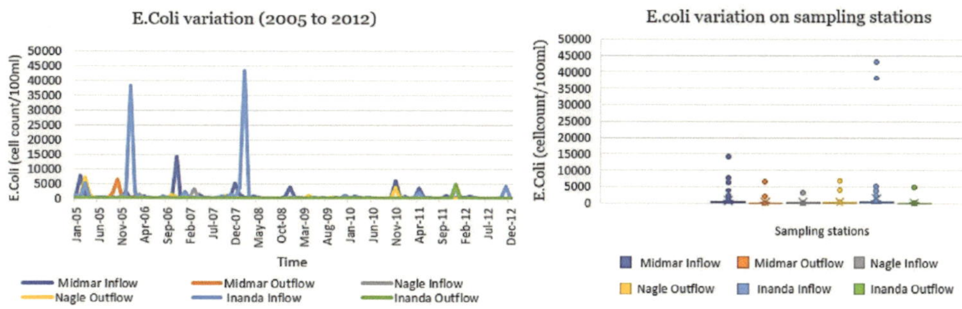

Figure 11.
E. coli time series and box-plot.

Sample	Minimum	Maximum	Median	Mean	S-K	Sen's slope	Decision
Midmar inflow	6.000	14200.000	272.500	745.319	0.572	−0.435	No trend
Midmar outflow	0.000	6490.000	23.000	126.729	0.235	−0.143	No trend
Nagle inflow	2.000	3100.000	38.000	126.246	0.132	0	No trend
Nagle outflow	7.750	6870.000	100.000	274.299	0.037	−0.47	Decrease
Inanda inflow	9.000	43100.000	213.000	1319.056	<0.0001	−3.038	Decrease
Inanda outflow	0.000	4838.000	26.000	97.917	0.449	0.05	No trend

Table 9.
Descriptive statistics for E. coli.

Treatment Plant. It is, therefore, reasonable to argue that more chlorine might be needed at Wiggins in order to thoroughly disinfect the water. Poor treatment of water with *E. coli* could cause the spread of waterborne diseases.

4.10 Ammonia

Ammonia concentration among the six stations studied showed irregular cyclic pattern with peaks (**Figure 12**). Of concern is the high monthly mean recorded at Midmar Dam outflow station of 3.640 mg/L (**Table 10**) that exceeded guidelines. Such an observation is worrisome when considering the toxicity of ammonia to aquatic animals, which is related to pH and temperature. Concentration 0.06 mg/L could damage fish gills while those above 0.3 mg/L be lethal to fish [67]. The presence of ammonia at higher than geogenic levels is an important indicator of faecal pollution [68].

Except for Nagle Dam station, the other dam inflow stations generally exhibited ammonia levels comparable to the corresponding outflow stations. Regarding treatment, ammonia is known to react with chlorine during treatment to form chloramine [21]. Chloramine is, however, a relatively weak oxidant [21] but with a long-lasting residual effect. Taste and odour problems as well as decreased disinfection efficiency are expected if drinking water containing more than 0.2 mg of ammonia per litre is chlorinated [68]. This is because up to 68% of the chlorine may react with the ammonia and become unavailable for disinfection [68]. Regarding the observed levels of ammonia in the data analysed, it can be argued that chlorine dosage at Wiggins Water Treatment Plant is more influenced by ammonia compared to Durban Heights Plant. The presence of elevated ammonia levels in raw water is also reported to interfere with the operation of manganese-removal filters as it may increase oxygen requirement due to nitrification and this may result in mouldy,

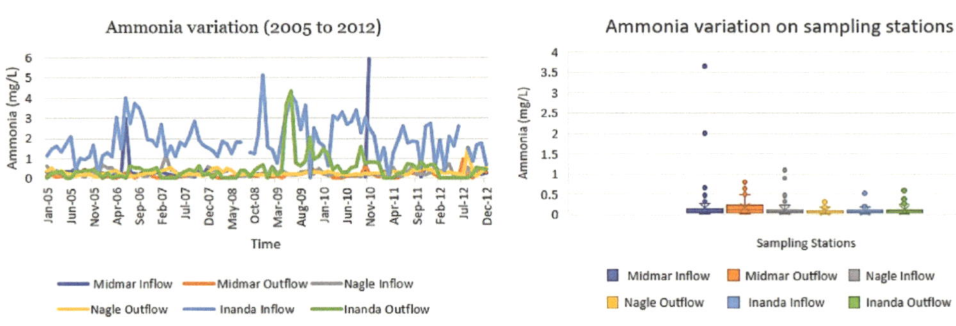

Figure 12.
Ammonia time series and box-plot.

Sample	Minimum	Maximum	Median	Mean	S-K	Sen's slope	Decision
Midmar inflow	0.020	3.640	0.070	0.162	1	−0.065	No trend
Midmar outflow	0.020	0.910	0.100	0.165	0.32	−0.08	No trend
Nagle inflow	0.010	1.100	0.060	0.100	0.004	−0.389	Decrease
Nagle outflow	0.020	0.400	0.050	0.075	0.074	−0.241	No trend
Inanda inflow	0.008	0.510	0.070	0.082	0.491	−0.234	No trend
Inanda outflow	0.008	0.600	0.080	0.105	0.347	−0.002	No trend

Table 10.
Descriptive statistics for Ammonia.

earthy-tasting water [68]. Furthermore, the presence of the ammonium cation in raw water may result in drinking-water containing nitrite as the result of catalytic action or the accidental colonisation of filters by ammonium-oxidising bacteria [69].

4.11 Electrical conductivity

Electrical Conductivity (EC) is the measure of dissolved ions or inorganic materials including calcium, bicarbonate, nitrogen, phosphorus, iron and sulphur. This parameter, which was used in Dzwairo, Otieno [5] as a surrogate for pollution, showed irregular variation with peaks and fluctuations (**Figure 13**). The distinct low level that was more pronounced in summer could be explained by an increased river flow and discharge, which tended to dilute the downstream watercourse pollution. However, all values observed (5.280 mS/m - 52.95 mS/m) were within the acceptable level for drinking water purposes, hence, are no cause for concern (**Table 11**).

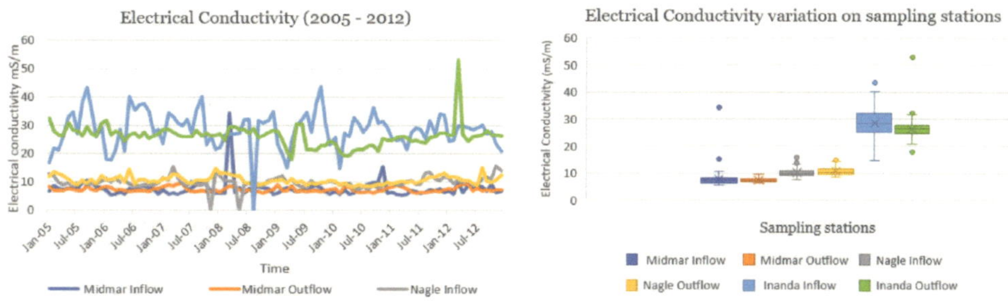

Figure 13.
Conductivity time series and the box-plot.

Sample	Minimum	Maximum	Median	Mean	S K	Sens slope	Decision
Midmar inflow	5.280	34.385	7.338	7.688	0.453	−0.003	No trend
Midmar outflow	5.990	9.400	7.080	7.339	0.248	−8.51E-04	No trend
Nagle inflow	7.360	15.680	9.765	10.100	0.179	0.004	No trend
Nagle outflow	8.430	14.915	10.330	10.622	0.045	−0.01	Decrease
Inanda inflow	14.740	43.600	28.600	28.644	0.278	−0.022	No trend
Inanda outflow	17.830	52.950	26.400	26.210	0.003	−0.028	Decrease

Table 11.
Descriptive statistics for Conductivity.

Previous studies have also reported uMngeni River as a low conductivity river. Graham [48], using data from 1990 to 1999, also noted that Nagle Dam had a daily average of 8.9 mS/m, which is slightly lower than the 10.361 mS/m observed in this study. This then implies that the concentration of dissolved ions at Nagle Dam has increased over time. As presumed, the box and whisker plot show that conductivity increased towards downstream of the uMngeni Basin. Regarding treatment, the relatively high conductivity observed at Inanda Outflow suggests that conductivity could be driving coagulant dosage more at Wiggins compared to Durban Heights. In an ecosystem, the combination of high EC coupled with high temperatures increases the toxicity of metals. At low temperatures EC tends to be low, which reduces the mobilisation of metals bound in sediments [70].

5. Conclusion

In a semi-arid country like South Africa, it is pertinent that the country's water resources be developed and managed for the benefit of the whole. Effective management of a basin such as uMngeni requires a sound understanding of the sources and status of water pollution as well as early detection of water quality changes. This will assist in strategising effective methods for combating pollution, making the water fit for various purposes including potable water production.

From the foregoing observations, it can be concluded that for the period under consideration, the quality of water along uMngeni River showed characteristics that favoured eutrophication with respect to turbidity, chloride, algae, conductivity and nitrate levels. This is more pronounced when comparing among the three inflow stations. Drinking water should be free from colour, turbidity, odour, and microbes in order to make it safe and aesthetically pleasant. Regarding the high *E. coli* levels that were observed at all stations, it could be concluded that the water was not fit for drinking purposes without at least disinfection. The high turbidity recorded at all stations suggests that the raw water was aesthetically unpleasant for drinking purposes.

The spatial analysis shows a downstream deterioration of river water quality for mineral salts namely chloride and conductivity. However, the quality of water was still acceptable for other uses such as irrigation, freshwater ecosystem habitat, and industrial use, which require less stringent guidelines than the standards for potable use. While, the quality of uMngeni River water was acceptable for agricultural purpose, it is important for the management and communities to implement effective measures for controlling agricultural practices that may significantly affect water quality such as stream bank cultivation and intensive fertiliser, which may be washed into the river. This is also the case with putting measures to control sewage bursts and open defecation, which have made the water unfit for human consumption. Future studies need to look at in-depth analysis of the ecological effectiveness of the temporal and spatial variation of the water quality in the basin.

Acknowledgements

The authors gratefully acknowledge Durban University of Technology for hosting and funding the main author during his Master's degree study.

Author details

Innocent Rangeti[1] and Bloodless (Rimuka) Dzwairo[2]*

1 Faculty of Health Sciences, Department of Community Health Studies, Durban University of Technology, Durban, South Africa

2 Durban University of Technology, Civil Engineering Midlands, Imbali, South Africa

*Address all correspondence to: bloodlessd@dut.ac.za

References

[1] Dzwairo B, Otieno FAO, Ochieng' GM. Making a case for systems thinking approach to integrated water resources management (IWRM). International Journal of Water Resources and Environmental Engineering. 2010;1(5): 107-13.

[2] Cullis JD, Rossouw N, Du Toit G, Petrie D, Wolfaardt G, De Clercq W, et al. Economic risks due to declining water quality in the Breede River catchment. Water SA. 2018;44(3): 464-73.

[3] Rangeti I, Dzwairo B, Barratt GJ, Otieno FAO. Ecosystem-specific water quality indices. African Journal of Aquatic Science. 2015;40(3):227-34.

[4] Dzwairo B, Otieno F, Ochieng G, Bezuidenhout J. Raw water quality weight factors: Vaal basin, South Africa. Water Science and Technology. 2012;66(5):1061-8.

[5] Dzwairo B, Otieno FAO, Ochieng' GM. Incorporating surface raw water quality into the cost chain for water services: Vaal catchment, South Africa. Research Journal of Chemistry and Environment. 2010;14(1):29-35.

[6] Kumar N, Singh SK, Singh VG, Dzwairo B. Investigation of impacts of land use/land cover change on water availability of Tons River Basin, Madhya Pradesh, India. Modeling Earth Systems and Environment. 2018;4(1):295-310.

[7] van Leeuwen FXR. Safe drinking water: the toxicologist's approach. Food and Chemical Toxicology. 2000;38: S51-S8.

[8] Dzwairo B, Ochieng GM, Letsoalo ME, Otieno FA. Data Reduction for Water Quality Modelling, Vaal Basin. In: Leite EP, editor. Scientific and Engineering Applications Using Matlab. Croatia: Intech; 2011. p. 71-86.

[9] Huang H, Wang Z, Xia F, Shang X, Liu Y, Zhang M, et al. Water quality trend and change-point analyses using integration of locally weighted polynomial regression and segmented regression. Environmental Science and Pollution Research. 2017;24(18):15827-37.

[10] Gonzales Inca C, Lepistö A, Huttula T. Trend detection in water-quality and load time-series from agricultural catchments of Yläneenjoki and Pyhäjoki, SW Finland. Boreal Environment Research. 2016;21:161-80.

[11] Reckhow K, Kepford K, Hicks WW. Methods for the analysis of lake water quality trends: EPA; 1993.

[12] Dallas HF, Pay JA, Reynolds EG. The effects of water quality variables on riverine biotas. Pretoria: Water Research Commission (WRC) 1994. Contract No.: Report 351.

[13] Dzwairo B, Otieno FA. Class frequency distribution for a surface raw water quality index in the Vaal Basin. Water SA. 2014;40(2):337-44.

[14] Palmer MA, Richardson DC. Provisioning services: A focus on fresh water. The Princeton Guide to Ecology Princeton University Press, Princeton, NJ. 2009:625-33.

[15] Dzwairo B, Hoko Z, Love D, Guzha E. Assessment of the impacts of pit latrines on groundwater quality in rural areas: A case study from Marondera district, Zimbabwe. Physics and Chemistry of the Earth, Parts A/B/C. 2006;31(15-16):779-88.

[16] Hoko Z, Dzwairo B, Sanyanga RA, Neseni N, Guzha E. A preliminary assessment of the gender sensitivity and health risk potential of ecological sanitation (ecosan) in Marondera Rural District, Zimbabwe. Journal of

Sustainable Development in Africa. 2010;12(1).

[17] Dzwairo B. Multi-date trends in groundwater pollution from pit latrines. Journal of Water Sanitation and Hygiene for Development. 2018:washdev2018177.

[18] Khan SS, Tareen H, Jabeen U, Mengal F, Masood Z, Ahmed S, et al., editors. Quality assessment of drinking water from the different colonies of Quetta city, Pakistan according to WHO Standards. Biological Forum; 2015: Research Trend.

[19] Pulles RW, Heat R, Howod M. A manual to assess manage the 06 gold mining operations on the surface water environmental. Pretoria: Water Research Commission (WRC); 1996. Contract No.: WRC Report No TT79/96.

[20] Otieno F, Ochieng G. Water management tools as a means of averting a possible water scarcity in South Africa by the year 2025. Water SA. 2007;30(5):120-4.

[21] Dzwairo B. Modelling raw water quality variability in order to predict cost of water treatment [DTech thesis]. Pretoria: Tshwane University of Technology; 2011.

[22] Gebremedhin SK. Analysis and Prediction of Chemical Treatment Cost of Potable Water in the Upper and Middle Vaal Water Management Areas [Master of Science in Agriculture]. Pietermaritzburg Campus: University of KwaZulu-Natal; 2009.

[23] Dennison DB, Lyne MC. Analysis and prediction of water treatment costs at the DV Harris Plant in Umgeni Catchment Area. Agrekon. 1997;36:40.

[24] Dzwairo B, Otieno FAO. Integrating quality and cost of surface raw water: Upper and Middle Vaal Water Management Areas South Africa. Water Science and Technology: Water Supply 2010;10(2):201-7.

[25] Voortman WJ, Reddy CD. Package Water Treatment Plant Selection,. South Africa: WRC; 1997. Contract No.: 450/1/97.

[26] Dearmont D, McCarl BA, Tolman DA. Costs of water treatment due to diminished water quality: a case study in Texas. Water Resources Research. 1998;34(4):849-54.

[27] uMgeni_Water_Amanzi. Corporate Profile 2020 [Available from: https://www.umgeni.co.za/about/].

[28] KZN_Treasury. Province of KwaZulu-Natal Provincial Budget–2019/20 2019 [Available from: http://www.treasury.gov.za/documents/provincial%20budget/2019/4.%20Guide%20to%20the%20Budget/KwaZulu-Natal%20-%202019%20Budget%20Summary.pdf].

[29] DWAF. uMngeni River and Neighbouring Rivers and Streams: State of the Rivers Report Pretoria2003 [Available from: http://www.dwaf.gov.za/iwqs/rhp/state_of_rivers/state_of_umngeni_02/umngeni.html].

[30] Roos J, Hassan F, Coleman T. Water Reconciliation Strategy Study of the Kwazulu Natal Coastal Metropplitan Areas. Water Quality Review Report. Department of Water Affairs and Forestry (DWAF); 2009. Contract No.: PWMA 11/000/00/2609.

[31] Deventer Rv. Impact of land use on water quality and aquatic ecosystem health of stream network in the upper uMngeni catchment feeding Midmar Dam, KwaZulu-Natal, South Africa [Masters]. Pietermaritzburg: University of KwaZulu-Natal; 2012.

[32] Olaniran A, Naidoo K, Pillay B. Assessment of physico-chemical qualities and heavy metal concentrations of Umgeni and Umdloti Rivers in Durban, South Africa. Environmental monitoring and assessment. 2013;186.

[33] Meals DW, Spooner J, Dressing SA, Harcum JB. Statistical Analysis for Monotonic Trends 2011 [Available from: https://www.epa.gov/polluted-runoff-nonpoint-source-pollution/nonpoint-source-monitoring-technical-notes].

[34] Morton R, Henderson BL. Estimation of nonlinear trends in water quality: an improved approach using generalized additive models. Water Resources Research. 2008;44(7).

[35] Helsel DR, Hirsch RM. Statistical Methods in Water Resources. Amsterdam, Netherlands: Elsevier Science Publisher B.V; 1992.

[36] Hirsch RM, Slack JR, Smith RA. Techniques of trend analysis for monthly water quality data. Water resources research. 1982;18(1):107-21.

[37] Antonopoulos VZ, Papamichail DM, Mitsiou KA. Statistical and trend analysis of water quality and quantity data for the Strymon River in Greece. Hydrology and Earth System Sciences. 2001;5(4):679-91.

[38] Hirsch RM, Slack JR, Smith RA. Techniques of trend analysis for monthly water quality analysis. Water Resource Research 1982;18 No 1:107-21.

[39] Darken PF. Testing for changes in trend in water quality data [Doctorate]. Blacksburg, Virginia: Virginia Polytechnic Institute and State University; 1999.

[40] DWAF. South African Water Quality Guidelines (second edition). Volume 1: Domestic Use. In: Forestry DoWAa, editor. Pretoria: Department of Water Affairs and Forestry; 1996. p. 214.

[41] Alkhatib E, Peters R. Wet weather impact on trihalomethane formation potential in tributaries to drinking water reservoirs. Environmental monitoring and assessment. 2008;139(1-3):173-81.

[42] WHO. Guidelines for Drinking Water quality. Geneva: World Health Organisation; 2004.

[43] Mann AG, Tam CC, Higgins CD, Rodrigues LC. The association between drinking water turbidity and gastro-intestinal illness: a systematic review. BMC public health. 2007;7(1):1-7.

[44] DWAF. South African Water Quality Guidelines: Volume 1: Domestic Water Use. Pretoria, South Africa: Department of Water Affiars and Forestry (DWAF); 1996.

[45] Kuutondokwa S. A dissertation submitted in partial fulfilment of the requirements for the degree of Masters in Intergrated Water Resources Management (IWRM). 2012.

[46] Pernitsky D, Edzwald J. Solubility of polyaluminium coagulants. Journal of Water Supply: Research and Technology - AQUA. 2003;52:395-406.

[47] Dzwairo B, Otieno F. Chemical pollution assessment and prioritisation model for the Upper and Middle Vaal water management areas of South Africa. IWA Journal of Water and Health. 2014; 12(4):14.

[48] Graham P. Modelling the water quality in dams within the Umgeni Water operational area with emphasis on algal relations [Philosophiae Doctor]: North West University; 2004.

[49] DWAF. South African Water Quality Guidelines, Vol 7: Aquatic Ecosystems. Pretoria: Department of Water Affairs and Forestry; 1996.

[50] Mesner N, Geiger J. Understanding Your Watershed: pH. Water Quality Extension, Utah State University. 2010.

[51] Dzwairo B. Assessment of the Impacts of Pit Latrines on Groundwater Quality in Rural Areas: a case study from Marondera District, Zimbabwe [MSc thesis]. Harare: University of Zimbabwe; 2005.

[52] Blignaut JN, Marais C, Turpie J. Determining a charge for the clearing

of invasive alien plant species (IAPs) to augment water supply in South Africa. Water Sa. 2007;33(1).

[53] Adriano DC. Trace Elements in Terrestrial Environments: Biogeochemistry, Bioavailability, and Risks of Metals: Springer New York; 2001.

[54] Dallas HF, Day JA. The effect of water quality variables on Aquatic Ecosystems: A Review. Pretoria, South Africa: Water Research Commission; 2004. Contract No.: TT224/04.

[55] Hanson A, Cleasby J. The effects of temperature flocculation: Fluid dynamics. . Journal of America Water Works Association. 1990;82(1):56-72.

[56] Veenstra JN, Schnoor JL. Seasonal-Variations in Trihalomethane Levels in an Iowa River Water-Supply. Journal American Water Works Association. 1980;72(10):583.

[57] Alkorta I, Epelde L, Garbisu C. Environmental parameters altered by climate change affect the activity of soil microorganisms involved in bioremediation. FEMS microbiology letters. 2017;364(19).

[58] Amiri M, Moghadasi J. The effect of temperature on calcium carbonate scale formation in Iranian oil reservoirs using OLI ScaleChem software. Petroleum science and technology. 2012;30(5):453-66.

[59] DWAF. South African Water Quality Guidelines. Volume 7: Aquatic Ecosystems. In: (DWAF) DoWAaF, editor. Pretoria, South Africa: Department of Water Affairs and Forestry (DWAF); 1996.

[60] Waston IM, Robinson JO, Burke V, Grace M. Invasive of Aeromonas spp. In relation to biotype virulence factors and clinical features. J Clin Microbiol. 1985;22(1):48-51.

[61] Manassaram DM, Backer LC, Moll DM. A review of nitrates in drinking water: maternal exposure and adverse reproductive and developmental outcomes. Environmental Health Perspectives. 2006;114(3):320-7.

[62] Schullehner J, Stayner L, Hansen B. Nitrate, nitrite, and ammonium variability in drinking water distribution systems. International Journal of Environmental Research and Public Health. 2017;14(3):276.

[63] Furness HD, Richard WN, editors. River Basin Management -Its role in protecting water quality South Africa: eWISA; 1987.

[64] Singh A, Lin J. Microbiological, coliphages and physico-chemical assessments of the Umgeni River, South Africa. International Journal of Environmental Health Research. 2015;25(1):33-51.

[65] Bartosova A, Strömqvist J, Arheimer B. Source to sea: Tracing causes of water quality problems and evaluating potential mitigation options. Case study of uMngeni River, south africa2017 07/10/2020 [cited 2020 07/10/2020]. Available from: https://www.smhi.se/polopoly_fs/1.117090!/SOURCE%20TO%20SEA%20-%20TRACING%20CAUSES%20OF%20WATER%20QUALITY%20PROBLEMS%20AND%20EVALUATING%20POTENTIAL%20MITIGATION%20OPTIONS.%20CASE%20STUDY%20OF%20UMNGENI%20RIVER,%20SOUTH%20AFRICA.pdf.

[66] Karar E. Promising water resources management approaches in the drinking water supply and sanitation sector: Evaluation document for the Mgeni Catchment Management Plan1997 07/10/2020 [cited 2020 07/10/2020]:[230 p.]. Available from: https://www.ircwash.org/sites/default/files/824-ZA-16276.pdf.

[67] Emerson K, Russo RC, Lund RE, Thurston RV. Aqueous ammonia equilibrium calculations: effect of pH and temperature. Journal of the Fisheries Board of Canada. 1975;32(12):2379-83.

[68] WHO. Ammonia in Drinking-water. Background document for development of WHO Guidelines for Drinking-water Quality. WHO/SDE/WSH/03.04/01. Originally published in Guidelines for drinking-water quality, 2nd ed. Vol. 2. Health criteria and other supporting information. World Health Organization, Geneva, 1996. Geneva: WHO Library Cataloguing-in-Publication Data; 2003. p. 9.

[69] WHO. Nitrate and nitrite in drinking-water. Background document for development of WHO Guidelines for Drinking-water Quality. WHO/SDE/WSH/07.01/16/Rev/1. Rev/1: Revisions indicated with a vertical line in the left margin. Geneva: WHO Library Cataloguing-in-Publication Data; 2011. p. 31.

[70] Li H, Shi A, Li M, Zhang X. Effect of pH, Temperature, Dissolved Oxygen, and Flow Rate of Overlying Water on Heavy Metals Release from Storm Sewer Sediments. Journal of Chemistry. 2013;2013:434012.

Chapter 5

Interpretation of Water Quality Data in uMngeni Basin (South Africa) Using Multivariate Techniques

Innocent Rangeti and Bloodless Dzwairo

Abstract

The major challenge with regular water quality monitoring programmes is making sense of the large and complex physico-chemical data-sets that are generated in a comparatively short period of time. Consequentially, this presents difficulties for water management practitioners who are expected to make informed decisions based on information extracted from the large data-sets. In addition, the nonlinear nature of water quality data-sets often makes it difficult to interpret the spatio-temporal variations. These reasons necessitated the need for effective methods of interpreting water quality results and drawing meaningful conclusions. Hence, this study applied multivariate techniques, namely Cluster Analysis and Principal Component Analysis, to interpret eight-year (2005–2012) water quality data that was generated from a monitoring exercise at six stations in uMngeni Basin, South Africa. The principal components extracted with eigenvalues of greater than 1 were interpreted while considering the pollution issues in the basin. These extracted components explain 67–76% of the water quality variation among the stations. The derived significant parameters suggest that uMngeni Basin was mainly affected by the catchment's geological processes, surface runoff, domestic sewage effluent, seasonal variation and agricultural waste. Cluster Analysis grouped the sampling six stations into two clusters namely heavy (B) or low (A), based on the degree of pollution. Cluster A mainly consists of water sampling stations that were located in the outflow of the dam (NDO, IDO, MDO and NDI) and its water can be described as of fairly good quality due to dam retention and attenuation effects. Cluster B mainly consist of dam inflow water sampling stations (MDI and IDI), which can be described as polluted if compared to cluster A. The poor quality water observed at Cluster B sampling stations could be attributed to natural and anthropogenic activities through point source and runoff. The findings could assist in determining an appropriate set of water quality parameters that would indicate variation of water quality in the basin, with minimum loss of information. It is, therefore, recommended that this approach be used to assist decision-makers regarding strategies for minimising catchment pollution.

Keywords: cluster analysis, multivariate technique, principal component analysis, uMngeni basin, water quality

1. Introduction

Water pollution is a global challenge undermining economic growth, health of millions of people as well as the physical status of the environment in both developed and developing countries. The current global water scarcity challenge is not only related to inadequacy in terms of quantity but also related to the progressive deterioration of quality making water unfit for some given uses such as potability. The deterioration of water quality is attributed to both natural (precipitation rate, weathering processes, soil erosion, etc) and anthropogenic (urban, industrial, agricultural activities, etc) factors. Seasonal variations in precipitation, surface run-off, ground water flow, interception and abstraction strongly affect the river discharge and the concentrations of water pollutants in a basin [1]. The effect of contaminant on water depends upon the characteristics of the water itself as well as quantity and characteristics of the contaminant.

Water pollutants, which are usually introduced through surface runoff or direct discharge, may in higher concentration result in rivers failing to provide adequate attenuation of pollutants, resulting in catchments failing to meet minimum compliance of quality for various uses such as potable water production. Furthermore, water quality deterioration is often a slow process not readily noticeable due attenuation effects until an apparent change occurs. Such situations are being exacerbated by the rapid increase in the demand for freshwater in many countries including South Africa. In view of the limited quantity of freshwater resources worldwide and the effect of anthropogenic activities, protection of these resources has become a priority [2–4]. It has, therefore, become imperative to monitor the quality of water in freshwater systems in order to prevent its further deterioration and thus ultimately ensure its continuous availability in a quality that meets various uses including potable water production. Pollutants in water can cause acute or chronic illness in humans especially when polluted water is consumed or when sewage is used to irrigate vegetables meant for human consumption. In specific cases this has resulted in loss of lives. For example, as at 2015, the bacterium *Vibrio cholerae* caused between 1.3 to 4.0 million infections and 21,000 to 143,000 deaths worldwide [5].

With concern of the detrimental effects of pollution, various agencies have been monitoring the quality of raw water within the uMngeni, a 232 km river that is then treated to serve almost 3.8 million people within and around Durban and Pietermaritzburg (South Africa) with potable water [6–10]. The primary objectives of such monitoring exercise have been to identify water quality problems, describe the spatio-temporal water quality trends, determine fitness compliance for specific uses and develop monitoring tools such as water quality indices for enhancing information dissemination. Although such monitoring programs are crucial to a better knowledge of hydrology and pollution problems in catchments such as uMngeni Basin, they tend to produce large amounts of complicated data-sets of various water parameters. The data-sets are often difficult to analyse and extract meaningful information and this makes it difficult to keep the public informed, who are the custodian of the resource. By keeping the public updated, it makes them more participatory in policy formulation and decision making regarding protection of the water resource [11, 12].

The classification and interpretation of monitoring stations are the most important steps in the assessment of water quality. Numerous studies have confirmed multivariate statistical techniques (cluster analysis, principal component analysis, factor analysis and discriminant analysis) as excellent tools for exploring and presenting the bulk and complex water quality data-sets [13–15]. These techniques allow for the determination of spatio-temporal water quality variability, classification of sampling stations and the identification of pollution sources [15–18]. Furthermore, by eliminating subjective assumptions, multivariate techniques tend to reduce biasness when selecting parameters for developing tools such as water quality index. This ultimately assists in improving the accuracy of such monitoring tool. Nevertheless, the selection of a multivariate technique to apply depends on the nature of data-set and research objectives. While there are a number of multivariate techniques, studies have extensively applied the Principal Component Analysis (PCA) and Cluster Analysis (CA) due to their suitability in extracting information on various situations [15, 19–22].

The application of principal component analysis (PCA) for the interpretation of a large and complex volume of data offers a better understanding of water quality, the ecological status of the basin being studied, while also allowing for the identification of possible factors/sources that influence the surface water systems [16]. Principal Component Analysis (PCA) aims to find combinations for certain variables to determine indices which describe the variation in the data while retaining as much information as possible. This reduction is achieved by transforming original variables into a new set of variables, known as principal components (PCs). These PCs, which are uncorrelated with the first few, retain most of the variation present in the original variables. The PCA technique transforms original variables into new uncorrelated variable known as principal components (PCs) [23–25]. The derived few variables can be used to provide a meaningful description of the entire data-set with a minimal loss of original information. The eigenvalues indicate the significance of each PC and a greater value, indicating the parameter's importance [26]. Correlation of PCs and original variables are given by the loadings [27]. While loadings reflect the relative importance of a variable within the component, it should be highlighted that these values does not show the importance of the component itself [28].

The PCA has been successfully applied on hydrogeological and hydrogeochemical studies. The application of PCA by Razmkhah, Abrishamchi [29] distinguished the anthropogenic and natural polluting activities along Jairood River in Iran. The results identified 5 factors which explained 85% the variation in water quality. Mazlum, Ozer [30] applied the PCA to determine factors causing water quality variability along a tributary, Porsuk, in Turkey. The study identified four PCs which explained 70% of the total water quality variance. The factors were related to the discharge of domestic wastewater, nitrification, industrial wastewater and the seasonal effect. Haag and Westrich [31] applied PCA to analyse the water quality along Neckar River in Germany based on ten parameters monitored from 1993 to 1998. The four principal components extracted accounting for 72% of total variance were interpreted as; (i) dilution by high discharge (ii) biological activity, (iii) seasonal effects and (iv) wastewater impact [31].

The limitations with the PCA technique include ignoring the degree of data dispersion as well as a weakness in processing nonlinear data. The result of PCA can also be influenced by uneven sampling interval, missing values or observations below detection limits of analytical methods, which can be changed during the data collection period. It is thus important to treat water quality data before modelling in order to improve the accuracy [32].

On the other hand, cluster analysis (CA) is an unsupervised pattern recognition multivariate technique which group objects (e.g., water quality variables) based on either their similarities or dissimilarities [33, 34]. Its objective is to sort cases into groups or clusters, so that the degree of association is strong between members of the same cluster and weak between members of different clusters. Most studies have applied the hierarchical clustering (HC) technique to sequentially category objects [13, 15, 34, 35]. Based on the hierarchical CA, water quality characteristics of each sampling location can be classified depending on pollution level. The results of a HC analysis are displayed graphically using a tree diagram commonly known as a dendrogram [13, 14, 36, 37]. The technique firstly groups the objects according to similarity. These groups are further merged according to their similarities or dissimilarities and eventually merge into a single cluster as the similarity among the subgroups decreases. The cluster analysis approach offers a reliable classification of surface water making it possible to design a future spatial sampling strategy that is cost-effective, with reduced number of sampling sites without losing any significant information [38].

2. Study area and water monitoring stations

uMngeni Basin, the study area, is situated in KwaZulu-Natal (KZN) Province, which lies along the eastern seaboard of the Republic of South Africa (**Figure 1**). uMngeni River (the main river in the basin), at 232 km long, is the primary source of raw water, which is then treated to serve a population of almost 3.8 million (as at 2013), in and around Durban metro as well as the city of Pietermaritzburg (PMB).

Key activities that generate point and non-point pollution within the catchment include agriculture and animal faming while concentrated urban settlements provide a variety of supportive economic activities that generate solid and liquid waste. A consequence of the concentrated development in the catchment area has been the high levels of pollutants entering the water system, which are eventually flushed out to sea. The basin receives much of its rain in summer, with occasional snow falls in some of its high lying areas such as the Drakensberg Mountain [39]. The geology of uMngeni Basin varies from basalts, granites, sandstones, shale and tillites [40].

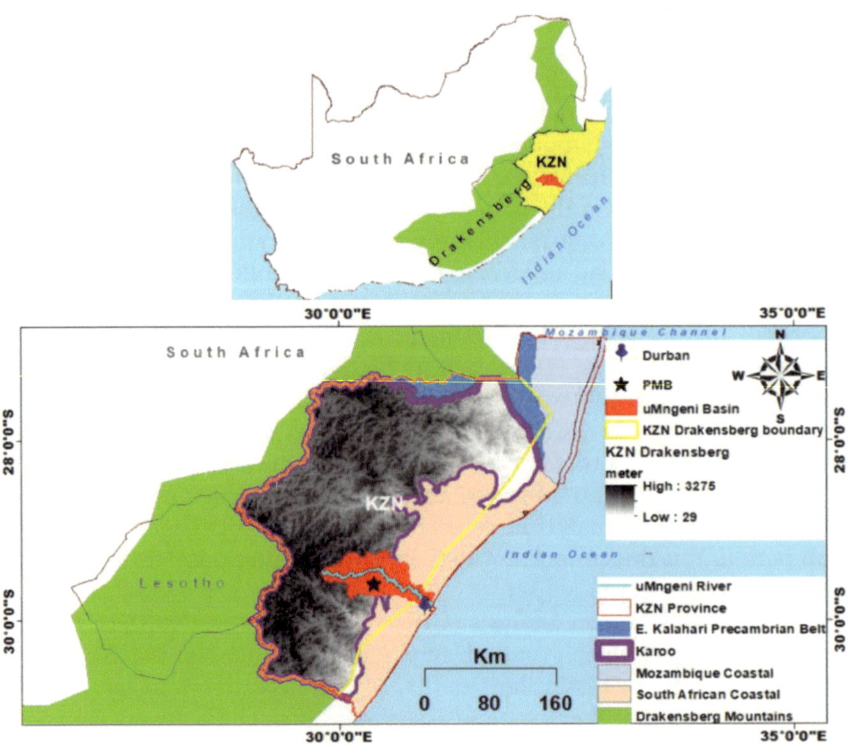

Figure 1.
uMngeni Basin, KZN Province, Drakensberg Mountains and South Africa.

About half of uMngeni Basin sits on top of the Karoo in the KZN part of the Drakensberg Mountains. The other potion extends east on top of the South African Coastal Plate (**Figure 2**).

The 2009 Landuse map (**Figure 2**) indicates that there are mixed activities, where cultivation and plantations are located predominantly from central to the north-west of the basin.

2.1 Water quality monitoring points considered

Six water quality monitoring points shown in **Figure 2**, namely Midmar Dam Inflow (MDI) *(Upstream point)*, Midmar Dam Outflow (MDO), Nagle Dam Inflow (NDI), Nagle Dam Outflow (NDO), Inanda Dam Inflow (IDI) and Inanda Dam Outflow (IDO) *(Downstream point)*, were considered in this study. The dam inflow stations were assumed to give a reflection of the pollution activities along the river course while the dam outflow stations were expected to depict the dilution and retention effects.

2.1.1 Methods and materials

Multivariate statistical methods have been widely applied in environmental data reduction and interpretation of multi-constituent chemical and physical biological measurements. These techniques have been applied to identify factors that influence water systems, to assist in reliable water resource management as well as determine rapid solutions for pollution problems [16, 41]. This study applied PCA and CA techniques to extract information from the raw data regarding the significant parameters influencing the variation of water quality at each of the six stations studied. The Kaiser-Meyer-Olkin (KMO), which test the sampling adequacy, was used to determine the suitability of water quality data for PCA analysis [42–44]. Kaiser [43] recommended 0.5 as a minimum (barely accepted), values between 0.7–0.8 acceptable, and values above 0.9 as depicting excellence. The current study employed the PCA technique to determine the most significant parameters that would explain the variation in water quality.

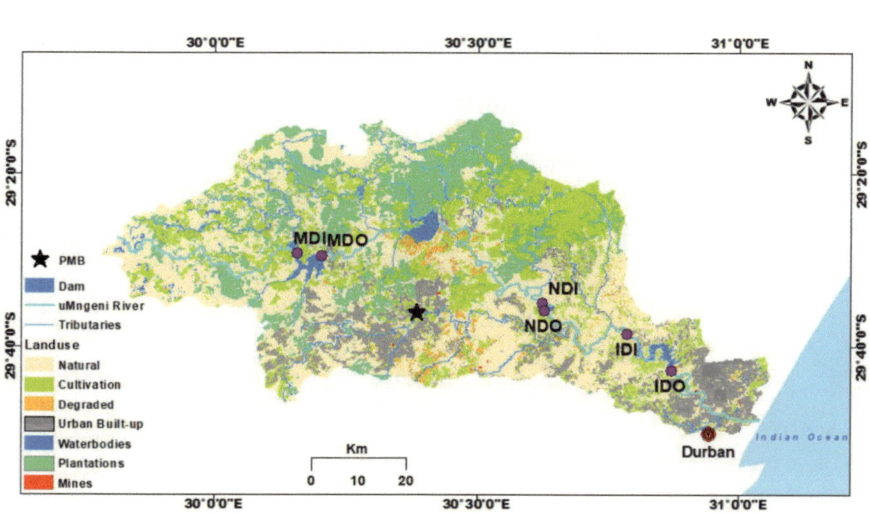

Figure 2.
Human settlements dominate from the central parts of the basin (PMB and its periphery) up to the Indian Ocean.

PCA is a very powerful multivariate statistical analysis technique used to reduce the dimensionality of a data set consisting multiple inter-related variables, while retaining data variability [45]. The technique extracts primary information representative of the typical characteristics of the water environment from a large amount of data and then represents it as a new set of independent variables of the principal component. PCA reduces the dimensionality of a multivariate data-set to a small number of independent principal components. Each principal component contains all the variable information, thus reducing the omission of information.

The PCA method is composed of five main operational steps, as follows:

1. The original data matrix is shown in Eq. 1:

$$X = \left(x_{ij}\right)_{n*p} = \begin{bmatrix} x_{11} & \cdots & x_{1p} \\ \vdots & \vdots & \vdots \\ x_{n1} & \cdots & x_{np} \end{bmatrix} \quad (1)$$

   where x_{ij} is the originally measured data, n represents the monitoring station, and p represents each water quality parameter.

2. Standardising the original data with Z-score standardisation formula to eliminate the impact of dimension (Eq. 2).

$$x_{ij}^{\star} = \left(x_{ij} - \overline{x_j}\right)/s_j, \quad (2)$$

   where x_{ij} is the standard variable, xj is the average value for jth indicator, and s_j is the standard deviation for the jth indicator.

3. Calculating the correlation coefficient matrix, R, with standardised data and determining the correlation between indicators (Eq. 3).

$$R = \left(r_{ij}\right)_{p\star p} = \frac{1}{n-1}\sum_{t=1}^{n} x_{ti}^{\star} * x_{tj}^{\star} \ (i, j = 1, 2\cdots, p) \quad (3)$$

4. Calculating the eigenvalues and eigenvectors of the correlation coefficient matrix, R, to determine the number of principal components. The eigenvalues of the correlation coefficient matrix, R, are represented by i (i = 1, 2 _ _ _ n) and their eigenvectors are ui ($U_i = U_{i1}, U_{i2}, \ldots\ldots\ldots U_{in}$) (i = 1, 2 _ _ _n). The value corresponds to the variance of the principal component, and the value of variance is positively correlated with the contribution rate of the principal components. Further, the cumulated contribution rate of the first m principal components should be more than 80%, which means that as explained in Eq. 4:

$$\sum_{i=1}^{m} \lambda_j / \sum_{i=1}^{n} \lambda_j \geq 0.80. \quad (4)$$

   The principal component is represented by Eq. 5.

$$F_i = u_{i1}x_1^{\star} + u_{i2}x_2^{\star} + \cdots + u_{in}x_n^{\star} \ (i = 1, 2, \cdots, n), \quad (5)$$

   where xi is the standardised indicator variable as shown in Eq. 6:

$$x_i^\star = (x_i - \overline{x_i})/s_i. \quad (6)$$

5. The obtained principal components are weighted and summed to obtain a comprehensive evaluation function, as shown in Eq. 7:

$$F = \frac{\lambda_1}{\lambda_1 + \lambda_2 + \ldots + \lambda_n} F_1 + \frac{\lambda_2}{\lambda_1 + \lambda_2 + \ldots + \lambda_n} F_2 + \ldots \frac{\lambda_n}{\lambda_1 + \lambda_2 + \ldots \lambda_n} F_n \quad (7)$$

Principal components with an eigenvalue greater than 1 were related to the major pollution sources in uMngeni Basin. Water quality parameters with loadings of greater than 0.5 (highlighted in bold in the results tables) were regarded as significantly influencing water quality variation in uMngeni Basin.

Thereafter, cluster analysis (CA) was applied to determine the spatial similarity of the six water sampling stations studied. The hierarchy cluster analysis was employed using the Ward's method with Euclidean distances as a measure of dissimilarity [37, 46]. The number of subgroups for analysis were determined by drawing a line across the dendrogram and examining the main clusters branching out beneath that line [47]. Determination of the subgroups for analysis was subjective based on available information regarding pollution activities the along uMngeni River.

Eight-year (2005–2012) water quality data-sets obtained from then Umgeni Water was used in this study. Since monitoring generally depends on the pollution problem at any given time and space, the number and type of parameters monitored at each of the stations varied. As the study was data-driven, a monthly median was used for in-depth analysis. The adoption of median instead of the mean was in consideration that the latter is normally influenced by the outliers which are common in water quality data-sets while the former is resistant. The period studied was determined in consideration of a criteria explained by Schertz, Alexander [48] and Lettenmaier, Conquest [49]. These studies reported that at least a five-year monthly data and two-year monthly data should be sufficient for a defensible monotonic and step-trend (abrupt shift) study, respectively.

3. Results

3.1 PCA analysis results

The Kaiser-Meyer-Olkin (KMO) results for the six stations ranged from 0.610 to 0.786 showing the fitness of the data-sets for PCA analysis. The component matrix tables shown in the different sections of the results only depict PCs with eigenvalue of greater than one (1). Only parameters with a correlation coefficient of great than 0.5 (highlighted in bold black) with its respective principal component were considered as significantly influencing water quality variability at any given station.

3.1.1 Midmar dam inflow (MDI)

Table 1 show the extracted seven PCs with eigenvalues of greater than 1 that explain 75% of the water quality variation at the Midmar Dam Inflow sampling station. While considering the high positive correlations of nutrient, metal ion and organic related parameters with component 1 (**Table 2**), it can be hypothesised that 20.6% of the water quality variation at this station is a result of both anthropogenic and natural processes. The nutrient and organic related parameters can be explained

Midmar dam inflow: total variance explained									
Component	Initial eigenvalues			Extraction sums of squared loadings			Rotation sums of squared loadings		
	Total	% of Variance	Cumulative %	Total	% of Variance	Cumulative %	Total	% of Variance	Cumulative %
1	5.740	24.955	24.955	5.740	24.955	24.955	4.740	20.607	20.607
2	5.002	21.750	46.705	5.002	21.750	46.705	3.621	15.742	36.349
3	1.903	8.272	54.977	1.903	8.272	54.977	3.011	13.090	49.440
4	1.325	5.760	60.737	1.325	5.760	60.737	1.787	7.771	57.211
5	1.165	5.065	65.802	1.165	5.065	65.802	1.412	6.139	63.350
6	1.103	4.795	70.598	1.103	4.795	70.598	1.390	6.043	69.393
7	1.011	4.394	74.992	1.011	4.394	74.992	1.288	5.598	74.992

Extraction Method: Principal Component Analysis.

Table 1.
Extracted values of the significant components at Midmar dam inflow (MDI).

Component Matrix[a]							
	Component						
	1	2	3	4	5	6	7
Potassium (K)	**0.890**	−0.094	−0.027	0.111	0.090	−0.054	−0.074
Sulphate (SO_4)	**0.852**	−0.027	−0.213	−0.078	−0.128	−0.018	0.072
Chloride (Cl)	**0.824**	−0.322	−0.169	−0.099	−0.129	−0.008	−0.071
Total Dissolved Solid	**0.743**	−0.199	−0.161	−0.197	−0.161	−0.001	−0.106
Nitrate (NO_3)	**0.666**	−0.243	−0.231	−0.097	−0.295	0.049	−0.014
Total Organic Carbon(TOC)	**0.662**	0.464	0.012	−0.053	0.074	0.003	−0.156
Escherichia coli (E. coli)	0.440	0.356	0.035	0.431	−0.153	0.135	−0.085
Suspended Solid	0.416	**0.741**	0.005	0.172	0.181	0.113	0.319
Iron (Fe)	0.401	**0.713**	−0.084	0.037	0.133	−0.101	0.152
Turbidity	0.400	**0.692**	0.064	0.209	0.190	0.141	0.404
Calcium (Ca)	**0.503**	**−0.687**	0.304	0.091	0.266	0.098	0.045
Magnesium (Mg)	**0.602**	**−0.681**	0.168	0.025	0.089	0.073	0.009
Sodium (Na)	**0.581**	**−0.666**	0.156	0.010	0.207	0.085	0.059
Alkalinity (Alk)	−0.036	**−0.632**	0.451	0.090	0.338	0.157	0.149
Total Phosphate	0.192	0.556	0.415	−0.092	−0.089	0.260	−0.284
Colour	0.246	0.545	0.225	−0.183	0.355	−0.198	0.026
Dissolved Oxygen (DO)	−0.048	−0.075	**−.591**	0.410	0.017	0.262	0.093
Temperature (Temp)	0.230	0.453	0.460	−0.383	0.017	−0.222	−0.137
Silicon (Si)	0.347	0.321	−0.379	−0.460	−0.026	0.059	0.121
pH	−0.059	−0.147	−0.433	−0.345	**0.550**	0.222	−.199
Ammonia (NH_3)	0.085	0.160	0.470	−0.089	−0.376	0.619	0.009
Conductivity	0.286	−0.324	0.225	0.099	−0.252	−0.528	0.339
Soluble Reactive Phosphate (SRP)	0.313	0.250	0.019	**0.513**	0.131	−0.223	**−0.599**

Extraction Method: Principal Component Analysis.
[a]7 Components extracted.
Bold: Significant contributors to the respective principal component in their respective there column.

Table 2.
The correlation among the parameters measured and the extracted significant components at MDI.

by piggery, dairy and maize farming activities surrounding Midmar Dam [6]. Animal manure enters surface water, both accidentally and deliberately, from households, villages, communal farms and feedlots. Without treatment, manure runoff tends to result in algae blooms which can lead to human health problems if consumed. The second component explained 15.7% (**Table 1**) of the total variance at MDI and showed a positive correlation to Suspended Solids (SS), iron (Fe) and turbidity (**Table 2**). Turbidity and suspended solids can be related to surface runoff from agricultural activities along uMngeni River whilst Iron (Fe) can be attributed to weathering processes.

Agriculture, a sector responsible for the usage of 70% of water being abstracted globally, plays a major role in water pollution [50, 51]. Runoff from agricultural activities such as agrochemicals, organic matter, drug residues, sediments and saline drainage into water bodies can lead to nutrient enrichment and eutrophication. The resultant water pollution poses a risk to aquatic ecosystems, human health and productive activities. Poor land management practises and deforestation can also explain the water quality variation at MDI. It is important for communities to practise improved land management through planting vegetation such as trees and plants to cover the ground. The negative relationship of metal ions with component 2 can be explained by the seasonality effect. Increased flow in the wet season turns to reduce the concentration of mineral salt content in a river system as a result of the dilution effect.

3.1.2 Midmar dam outflow (MDO)

The results in **Table 3** indicate that the first seven principal components with eigenvalues of greater than one (1) account for 73.7% of the total variance in the water-quality data set at Midmar Dam Outflow station. Component 1 which explains 27.1% of the total variance at MDO (**Table 3**) is mainly influenced by parameters related to human activities (turbidity and ammonia) as well as natural geological processes (silicon and calcium) (**Table 4**). Silicon is part of various essential plant minerals and it is released during weathering processes. Sodium and potassium which showed a moderate positive correlation to component 5 reflects the natural processes such as weathering.

3.1.3 Nagle dam inflow

The PCA technique identified five components, which cumulatively explained 67.4% of the total variance at NDI (**Table 5**). Component 1 which explained 24.7% of the total variance is significantly affected by parameters (highlighted in bold black) which normally originate from surface runoff of agriculture areas and effluent from wastewater treatment plants (**Table 6**). The pollution of water bodies when practicing agriculture is mainly due to fertiliser runoff after rainfall, nutrients (such as nitrogen) that percolate through the soil and contaminated groundwater, as well as sediment that is eroded from fields and washed into watercourses during and after rainfall. While studying the limnology of South Africa's major impoundments, Walmsely and Butty [52] described Nagle Dam as a phosphate limited oligotrophic system.

3.1.4 Nagle dam outflow

Seven significant components which explained 75% of the total variance were extracted at Nagle Dam Outflow station (**Table 7**). The first component which contributed 23% (**Table 8**) of the total variance is dominated with metal ions which can be related on natural geological processes such as weathering. The second component which explains 15.5% of the total variation is dominated by *E. coli*, suspended solids and turbidity (highlighted in bold black) (**Table 8**). These pollutants can be related to the discharge of sewage effluent and runoff from a community practicing open defecation.

3.1.5 Inanda dam inflow

At Inanda Dam Inflow station, seven components which explained 76% of the total variance were extracted (**Table 9**). Component 1 (**Table 10**) which is mainly metal ions and explains 11% of the total variance was comprised of metal ions (highlighted in bold black) which suggest that geological processes in the area could be significantly attributing to the water quality variation. The high positive correlation of chloride and component 1 also reflects the effect of anthropogenic pollutants

Midmar dam outflow: total variance explained									
Component	Initial eigenvalues			Extraction sums of squared loadings			Rotation sums of squared loadings		
	Total	% of Variance	Cumulative %	Total	% of Variance	Cumulative %	Total	% of Variance	Cumulative %
1	6.729	32.044	32.044	6.729	32.044	32.044	5.690	27.096	27.096
2	2.042	9.726	41.770	2.042	9.726	41.770	1.871	8.910	36.006
3	1.759	8.378	50.148	1.759	8.378	50.148	1.794	8.542	44.548
4	1.399	6.660	56.808	1.399	6.660	56.808	1.769	8.424	52.971
5	1.271	6.053	62.861	1.271	6.053	62.861	1.659	7.902	60.873
6	1.196	5.694	68.555	1.196	5.694	68.555	1.508	7.181	68.054
7	1.078	5.133	73.687	1.078	5.133	73.687	1.183	5.634	73.687

Extraction Method: Principal Component Analysis.

Table 3.
Extracted values of the significant components at Midmar dam outflow.

Component Matrix[a]							
	Component						
	1	2	3	4	5	6	7
Calcium (Ca)	**0.906**	−0.100	0.117	−0.076	0.134	−0.065	0.054
Conductivity	**0.869**	0.028	0.083	0.114	0.030	0.060	−0.050
Magnesium (Mg)	**0.856**	−0.131	0.109	−0.236	0.171	−0.045	0.039
Alkalinity	**0.821**	−0.208	0.103	−0.103	−0.092	0.080	0.075
Ammonia (NH_3)	**0.783**	−0.136	0.201	−0.051	−0.181	0.072	−0.036
Turbidity	**0.736**	0.442	−0.045	−0.171	−0.183	−0.146	0.062
Suspended Solids (SS)	**0.679**	**0.533**	−0.039	−0.187	−0.118	−0.163	0.079
Sulphates (SO_4)	**−0.663**	0.052	−0.267	0.088	0.314	0.107	0.315
Silicon (Si)	**0.613**	0.253	0.099	0.390	0.123	−0.164	0.086
Dissolved Oxygen (DO)	**−0.544**	0.260	0.062	−0.368	0.195	−0.143	−0.019
Nitrate (NO_3)	**0.519**	−0.334	0.068	0.130	−0.303	0.288	−0.061
Total Phosphate (TP)	0.205	**0.532**	−0.285	−0.332	0.016	0.291	−0.048
% NH3%	−0.247	0.292	**0.791**	0.065	0.097	0.314	0.154
pH	−0.336	0.463	**0.732**	0.005	0.061	0.103	0.130
Total Organic Carbon (TOC)	0.212	0.412	−0.169	0.459	−0.116	−0.355	0.306
Potassium (K)	0.399	0.046	−0.1608	0.308	**0.605**	−0.105	−0.160
Sodium (Na)	0.278	−0.360	0.245	−0.421	**0.533**	−0.148	−0.084
Chloride (Cl)	0.384	0.277	−0.303	−0.104	0.390	0.211	0.193
Temperature `C	0.320	−0.159	−0.057	0.437	0.209	0.558	0.160
SRP	0.063	0.434	−0.279	−0.095	−0.048	0.488	−0.443
Escherichia coli (E. coli)	−0.031	0.230	0.225	0.355	0.127	−0.195	**−0.718**

Extraction Method: Principal Component Analysis.
[a]7 Components extracted.
Bold: Significant contributors to the respective principal component in their respective there column.

Table 4.
The correlation among the parameters measured and the extracted significant components at MDO.

on this station. Both legal and illegal effluent discharges from industrial areas such as Willowton, are the predominant pollution sources that affect Inanda Dam [53]. In developing countries, 70 percent of industrial wastes are dumped untreated into waters, impacting on the usability of the water resource [54].

Given the high positive correlation coefficient of Soluble Reactive Phosphate (SRP), Suspended Solids (SS), Total Organic Carbon (TOC), Total Phosphate (TP) and turbidity on the component 2, it can be claimed that pollutants in this group which explained 9% of the variation in water quality are emanating from agricultural activities (**Table 10**). The negative correlation noted between dissolved oxygen and component 5 indicates deterioration in the water quality. Since component 5 also exhibited a positive correlation to temperature, it can be deduced that climatic conditions could explain the 7.7% variation noted at this station. Component 6 which is mainly influenced by *E. coli* suggest the effects of sewage effluent into the water system. This could be attributed to the effluent from DV Wastewater Works which treat domestic effluent from Pietermaritzburg.

Nagle dam inflow: total variance explained									
Component	Initial eigenvalues			Extraction sums of squared loadings			Rotation sums of squared loadings		
	Total	% of Variance	Cumulative %	Total	% of Variance	Cumulative %	Total	% of Variance	Cumulative %
1	4.960	30.998	30.998	4.960	30.998	30.998	3.962	24.760	24.760
2	2.164	13.523	44.521	2.164	13.523	44.521	2.073	12.957	37.717
3	1.429	8.934	53.455	1.429	8.934	53.455	1.973	12.332	50.049
4	1.161	7.257	60.713	1.161	7.257	60.713	1.558	9.736	59.785
5	1.074	6.710	67.422	1.074	6.710	67.422	1.222	7.638	67.422

Extraction Method: Principal Component Analysis.

Table 5.
Extracted values of the significant components at Nagle dam inflow (NDI).

Component Matrix[a]					
	Component				
	1	2	3	4	5
Turbidity	**0.822**	−0.253	−0.036	−0.214	0.052
Total Phosphate	**0.776**	−0.084	−0.350	0.307	0.068
Soluble Reactive Phosphate (SRP)	**0.757**	−0.207	−0.323	0.299	0.080
Suspended Solids (SS)	**0.741**	0.001	−0.213	−0.333	0.093
Total Organic Carbon (TOC)	**0.721**	0.099	−0.132	−0.410	0.152
Escherichia coli (E. coli)	**0.718**	−0.404	0.047	0.273	0.234
Nitrate (NO_3)	**0.677**	−0.278	0.337	−0.076	−0.251
Conductivity	**0.662**	0.020	0.383	−0.180	−0.336
Temperature	**0.566**	0.395	0.205	0.131	−0.325
% NH3%	0.363	**0.844**	−0.137	−0.088	0.152
pH	0.145	**0.819**	−0.094	0.010	0.228
Scenedesmus	0.160	0.336	−0.119	0.246	−0.283
Nitzschia	0.206	0.153	**0.711**	−0.130	0.087
Algal count cell	0.253	0.301	0.093	0.486	−0.334
Navicula	0.143	0.016	0.517	0.409	**0.612**
Ammonia- N (NH_3)	−0.051	−0.185	−0.106	0.183	−0.206

Extraction Method: Principal Component Analysis.
[a]5 Components extracted.
Bold: Significant contributors to the respective principal component in their respective there column.

Table 6.
The correlation among the parameters and the significant components at NDI.

Nagle dam outflow: total variance explained						
Component	Initial eigenvalues			Extraction sums of squared loadings		
	Total	% of Variance	Cumulative %	Total	% of Variance	Cumulative %
1	4.832	23.011	23.011	4.832	23.011	23.011
2	3.262	15.532	38.542	3.262	15.532	38.542
3	2.186	10.409	48.951	2.186	10.409	48.951
4	1.879	8.946	57.897	1.879	8.946	57.897
5	1.390	6.618	64.515	1.390	6.618	64.515
6	1.185	5.643	70.158	1.185	5.643	70.158
7	1.072	5.103	75.261	1.072	5.103	75.261

Extraction Method: Principal Component Analysis.

Table 7.
Extracted values of the significant components at Nagle dam outflow.

3.1.6 Inanda dam outflow

At Inanda Dam Outflow station, eight components explaining 75% of the total variance were extracted as depicted in **Table 11**. We hypothesised that pollutants in component 1 depicted in **Table 12** (11.8% and highlighted in bold black) were

Component Matrix[a]							
	Component						
	1	2	3	4	5	6	7
% NH3%	−.158	.051	.364	**.690**	.483	−.092	.164
Alkalinity	.342	−.099	.339	−.377	.102	.149	.346
Calcium (Ca)	**.650**	−.537	−.454	.154	.103	−.007	.030
Chloride (Cl)	**.673**	−.154	.499	−.032	.055	.037	−.281
Conductivity	**.782**	−.017	.430	−.134	.114	.024	−.039
Dissolved Oxygen (DO)	−.496	−.280	.224	−.228	.016	.255	−.213
Escherichia coli (E. coli)	.330	**.630**	−.300	−.071	.287	.227	−.175
Potassium (K)	**.707**	−.235	−.219	.259	−.269	−.213	.031
Magnesium (Mg)	**.774**	−.557	−.120	.036	.126	.087	.050
Sodium (Na)	.549	−.591	−.482	.257	.017	−.040	−.050
Ammonia (NH_3)	−.027	.091	−.381	.117	−.047	.453	.426
Nitrate (NO_3)	.498	.053	**.511**	.232	−.337	.048	−.408
pH	−.350	−.039	.181	**.671**	**.524**	−.032	−.052
Silicon (Si)	.442	.404	.255	−.162	.043	.011	.364
Sulphate (SO_4)	.326	−.363	.139	−.445	**.583**	.266	.000
Soluble Reactive Phosphate (SRP)	.056	.148	−.166	−.385	.302	**−.597**	−.028
Suspended Solids (SS)	.312	**.618**	−.238	.193	.046	.404	−.197
Temperature	**.506**	.429	.292	.184	−.183	−.177	.441
Total Organic Carbon	.489	.497	.031	.116	−.047	−.133	−.031
Total Phosphate (TP)	.127	.354	−.303	−.227	.283	−.347	−.130
Turbidity	.402	**.692**	−.282	.006	.084	.109	−.165

Extraction Method: Principal Component Analysis.
[a]7 Components extracted.

Table 8.
The correlation among the parameters and the significant components at NDO.

mainly contributed by metal ions which reflects the geology of a catchment area. The positive correlation of sulphate and component 1 reflects the effect of anthropogenic polluting activities. Component 3 (**Table 12**) is mainly attributable to agricultural pollutant sources due to moderate positive high correlations with turbidity, nitrate and suspended solids. It is most plausible to suggest that turbidity and suspended solids is a result of surface runoff due to rainfall. *Escherichia coli (E. coli)* which dominates in component 6 normally can be attributed to wastewater treatment plants in Pietermaritzburg.

3.2 Cluster analysis

Cluster analysis was used to detect similarities among the sampling stations in the study area. The dendrogram shows that the six sampling stations in the area studied could be grouped into two significant clusters (A and B) as illustrated by **Figure 3**. Such is the case of the relatively large linkage distance at which the two groups combine, which indicates the Euclidean distances [47]. Cluster A mainly

Inanda dam inflow: total variance explained									
Component	Initial eigenvalues			Extraction sums of squared loadings			Rotation sums of squared loadings		
	Total	% of Variance	Cumulative %	Total	% of Variance	Cumulative %	Total	% of Variance	Cumulative %
1	7.005	31.840	31.840	7.005	31.840	31.840	5.946	27.025	27.025
2	2.888	13.127	44.967	2.888	13.127	44.967	2.443	11.104	38.129
3	1.865	8.478	53.445	1.865	8.478	53.445	2.019	9.179	47.309
4	1.427	6.485	59.930	1.427	6.485	59.930	1.922	8.737	56.045
5	1.363	6.196	66.127	1.363	6.196	66.127	1.695	7.704	63.749
6	1.171	5.325	71.451	1.171	5.325	71.451	1.385	6.297	70.046
7	1.012	4.599	76.050	1.012	4.599	76.050	1.321	6.004	76.050
Extraction Method: Principal Component Analysis.									

Table 9.
Extracted values of the significant components at Inanda dam inflow.

Component Matrix[a]							
	Component						
	1	2	3	4	5	6	7
% NH3%	−0.155	0.269	**0.718**	0.193	0.209	−0.185	−0.170
Alkalinity	**0.746**	−0.186	0.182	−0.241	0.254	0.045	−0.065
Calcium (Ca)	**0.859**	0.215	−0.059	−0.247	0.030	0.093	0.063
Chloride (Cl)	**0.911**	0.140	0.089	−0.054	0.122	0.024	−0.090
Conductivity	**0.889**	0.169	0.009	−0.048	0.108	0.027	−0.096
Dissolved Oxygen (DO)	0.313	−0.033	0.128	0.464	**−0.640**	0.168	0.017
Escherichia coli	−0.297	0.400	−0.055	0.080	−0.122	**0.538**	−0.355
Flourine (F)	0.204	0.307	0.318	0.329	−0.067	0.236	−0.319
Potassium (K)	**.514**	0.496	−0.065	−0.122	−0.012	−0.240	0.116
Magnesium (Mg)	**0.734**	0.145	0.075	−0.161	0.074	0.411	0.176
Sodium (Na)	**0.882**	0.174	0.100	−0.017	0.045	0.009	−0.006
Ammonia (NH_3)	0.163	0.330	0.194	0.239	−0.195	0.012	**0.681**
Nitrate (NO_3)	0.493	0.194	−0.375	0.336	−0.191	−0.131	0.144
pH	−0.075	0.227	**0.717**	0.019	−0.260	−0.408	−0.102
Silicon (Si)	−0.581	0.216	0.178	0.174	0.322	0.192	0.306
Sulphate (SO_4^{2+})	**0.799**	.339	0.088	0.076	0.071	0.061	0.005
Soluble Reactive Phosphate (SRP)	0.086	**0.553**	−0.461	0.396	0.271	−0.278	−0.102
Suspended Solids	−.522	**0.569**	−0.048	−0.382	−0.219	0.006	0.108
Temperature	−0.506	0.187	0.303	0.088	**0.544**	0.184	0.215
Total Organic Carbon (TOC)	−0.190	**0.557**	0.036	**−0.516**	−0.192	−0.249	−0.049
Total Phosphate (TP)	−0.303	**0.645**	−0.351	0.229	0.227	−0.142	−0.158
Turbidity	−0.585	**0.620**	−0.028	−0.241	−0.184	0.328	−0.010

Extraction Method: Principal Component Analysis.
[a]7 Components extracted.
Bold: Significant contributors to the respective principal component in their respective there column.

Table 10.
The correlation among the parameters and the significant components at IDI.

consists of four sampling stations that were located mostly in the outflow of the river (NDO, IDO, MDO and NDI) while Cluster B mainly consist of two stations mainly dam inflow stations (MDI and IDI). Except for Nagle Dam Inflow, Cluster A basically comprised of dam outflow stations. These stations can be described as less polluted due to the dilution and retention effect. On the other hand, Cluster B composed of dam inflow stations. These stations can be described as more polluted as a result of activities along the river course. The PC results explained Section 4.1 of this chapter showed that poor agriculture practises resulting in runoff of agro-chemicals, organic matter, drug residues, sediments and saline drainage as well as sewage and industrial effluent discharges are key factors being reflected by the poor water quality results of the dam's inflow stations (Cluster B). These practices pose a risk to aquatic ecosystems, human health and productive activities. The significant presence of *E. coli*, suspended solids and turbidity in both Cluster A and B sampling stations indicates that raw water along uMngeni Basin is not fit for potable use before treatment.

Inanda dam outflow: total variance explained						
Component	**Initial eigenvalues**			**Extraction sums of squared loadings**		
	Total	**% of Variance**	**Cumulative %**	**Total**	**% of Variance**	**Cumulative %**
1	5.575	25.342	25.342	5.575	25.342	25.342
2	2.597	11.804	37.146	2.597	11.804	37.146
3	2.066	9.390	46.536	2.066	9.390	46.536
4	1.582	7.192	53.728	1.582	7.192	53.728
5	1.360	6.183	59.911	1.360	6.183	59.911
6	1.217	5.534	65.445	1.217	5.534	65.445
7	1.135	5.161	70.606	1.135	5.161	70.606
8	1.009	4.589	75.194	1.009	4.589	75.194
Extraction Method: Principal Component Analysis.						

Table 11.
Extracted values of the significant components at Inanda dam outflow.

Component Matrix[a]								
	Component							
	1	**2**	**3**	**4**	**5**	**6**	**7**	**8**
Magnesium (Mg)	**0.903**	−0.053	−0.074	−0.034	−0.141	0.023	0.013	−0.014
Sodium (Na)	**0.881**	0.007	−0.096	−0.078	−0.102	0.030	−0.107	−0.029
Calcium (Ca)	**0.877**	−0.033	0.056	0.108	−0.146	0.174	−0.189	0.000
Chloride (Cl)	**0.841**	−0.050	−0.031	0.029	0.071	0.170	0.257	0.001
Potassium (K)	**0.800**	−0.076	−0.093	−0.100	−0.138	−0.411	−0.045	−0.011
Conductivity	**0.671**	−0.096	−0.186	0.042	0.241	0.182	0.046	−0.131
Alkalinity	**0.664**	−0.334	−0.367	0.268	0.138	0.059	−0.009	0.158
Sulphate (SO_4)	**0.612**	0.374	0.080	−0.262	−0.118	0.055	0.095	−0.324
% NH3%	0.165	**0.816**	−0.177	−0.033	−0.212	−0.037	−0.011	0.376
pH	0.212	**0.781**	−0.046	−0.084	−0.295	0.039	−0.090	0.310
Temperature	0.027	**0.596**	−0.134	−0.120	0.347	−0.151	0.470	0.041
Turbidity	0.327	0.099	**0.785**	0.153	0.143	−0.057	0.028	0.046
Nitrate (NO_3)	−0.009	0.192	**0.721**	0.039	−0.279	0.294	−0.290	−0.197
Suspended Solids	0.325	0.347	**0.531**	0.121	0.317	−0.194	0.170	−0.178
Ammonia (NH_3)	0.271	−0.384	0.430	−0.071	−0.094	0.011	0.004	0.290
Total Phosphate (TP)	−0.008	−0.226	0.106	**−0.636**	−0.032	0.172	−0.051	0.217
Soluble Reactive Phosphate (SRP)	0.017	−0.366	0.232	**−0.605**	−0.017	−0.043	0.160	0.371
Flourine (F)	0.469	−0.136	−0.003	−.103	0.538	−0.089	−0.300	0.130
Dissolved Oxygen (DO)	0.041	−0.213	−0.176	0.398	−0.508	0.042	0.083	0.045
Escherichia coli (E coli)	−0.094	0.149	−0.026	0.171	0.282	**0.832**	0.139	0.199
Silicon (Si)	0.095	−0.283	0.327	0.458	−0.161	−0.146	0.524	0.300

Component Matrix[a]								
	Component							
	1	2	3	4	5	6	7	8
TOC	−0.047	0.169	0.064	0.407	0.283	−0.203	−0.496	0.363

Extraction Method: Principal Component Analysis.
[a]8 Components extracted.
Bold: Significant contributors to the respective principal component in their respective there column.

Table 12.
The correlation among the parameters and the significant components extracted at IDO.

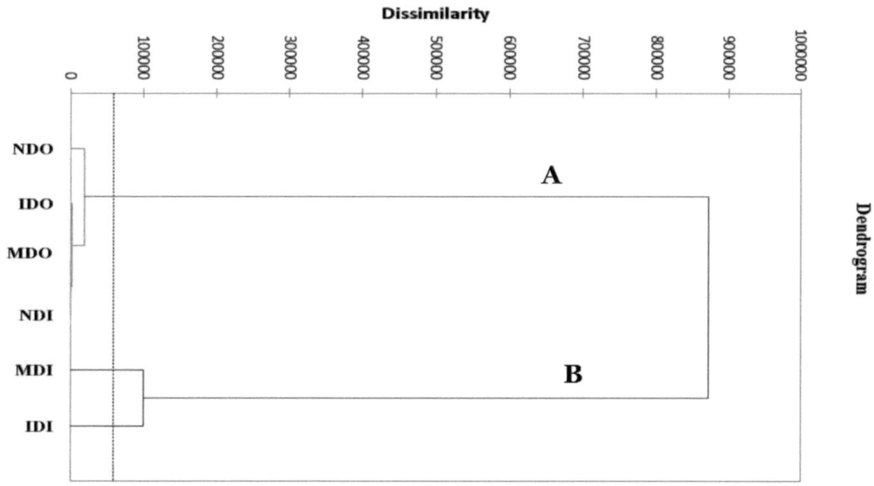

Figure 3.
Dendrogram of the stations along uMngeni basin.

4. Conclusions

Understanding the primary effects of anthropogenic activities and natural factors on river water quality is important in the study and efficient management of water resources. Hence, the PCA method assisted in the identification of significant parameters influencing water quality variations at the six stations studied in uMngeni Basin. The PCs extracted suggest that pollution sources along uMngeni Basin can be attributed to geological processes, sewage effluent, agricultural runoff and surface runoff pollutants. The findings could assist in reducing the number of parameters being monitored at any station and thus ultimately reducing the associated cost monitoring cost. It is recommended that, effluents be treated before discharge into the river. Additionally, it is recommended that buffer zone policies be enforced.

The result of the cluster analysis should also assist in categorising sampling sites according to pollution levels. Classification of sampling stations based on pollution level can assist in the designing of an optimal sampling strategy, which could reduce the number of sampling stations and associated costs. This study highlights the usefulness of multivariate statistical assessment such as PCA and CA in analysing complex databases, especially in the identification of pollution sources and to better comprehend the spatial and temporal variations for effective river water-quality management. It is worthwhile to conclude that PCA and CA are better tools for better understanding concealed information about parameter variance and datasets. The study recommends the application of PCA and CA for interpreting bulk surface water quality data-sets.

Acknowledgements

The authors gratefully acknowledge Durban University of Technology for hosting and funding the main author during his Master's Degree study.

Author details

Innocent Rangeti[1]* and Bloodless Dzwairo[2]

1 Department of Community Health Studies, Faculty of Health Sciences, Durban University of Technology, Durban, South Africa

2 Civil Engineering Midlands, Durban University of Technology, Imbali, South Africa

*Address all correspondence to: innoranger@gmail.com

References

[1] Vega M, Pardo R, Barrado E, Debán L. Assessment of seasonal and polluting effects on the quality of river water by exploratory data analysis. Water Research. 1998;**32**(12):3581-3592

[2] Dinar S. Water, security, conflict, and cooperation. SAIS Review. 2002; **22**(2):229-253

[3] Chinhanga JR. Impact of industrial effluent from an iron and steel company on the physico-chemical quality of Kwekwe River water in Zimbabwe. International Journal of Engineering, Science and Technology. 2010;**2**(7)

[4] Wei S, Wang Y, Lam JC, Zheng GJ, So M, Yueng LW, et al. Historical trends of organic pollutants in sediment cores from Hong Kong. Marine pollution bulletin. 2008;**57**(6):758-766

[5] Ali M, Nelson AR, Lopez AL, Sack DA. Updated global burden of cholera in endemic countries. PLoS neglected tropical diseases. 2015;**9**(6): e0003832

[6] Department of Water Affairs and Forestry. uMngeni River and Neighbouring Rivers and Streams. Forestry DoWAa: State of the Rivers Report Pretoria; 2003

[7] Umgeni Water. Umgeni Water Infrastructure Master Plan. (2012/2013–2042-2043). Pietremarizburg, South Africa: Umgeni Water. Palanning Services EaSSD: UW; 2012. p. 2012

[8] VanDerZel DW. Umgeni River Catchment Analysis. Water SA. 1975;1(2).

[9] Hemens J, Simpson DE, Warwick RJ. Nitrogen and phosphorus input to the Midmar dam. Natal. Water SA. 1977; **3**(4):193-201

[10] Statement UWUWP, Volume 2 F-YBP. Business Plan. 2012/13 to 2016/17. Pietermarizburg. Water U: Umgeni Water; 2012

[11] Couillard D, Lefebvre Y. Analysis of water indices. Journal of Environmental Management. 1985;**21**(2):161-179

[12] Wepener V, Cyrus DP, Vermeulen LA. O'Brien GC, Wade P. Development of a Water Quality Index for Estuarine Water Quality Management in South Africa. Commission WR, editor. Pretoria; 2006

[13] Yerel S, Ankara H. Application of multivariate statistical techniques in the assessment of water quality in Sakarya River, Turkey. Journal of the Geological Society of India. 2011;**78**(6):1-5

[14] Alkarkhi AM, Ahmad A, Easa A. Assessment of surface water quality of selected estuaries of Malaysia: Multivariate statistical techniques. Environmentalist. 2009;**29**(3):255-262

[15] Singh KP, Malik A, Mohan D, Sinha S. Multivariate statistical techniques for the evaluation of spatial and temporal variations in water quality of Gomti River (India)—A case study. Water Research. 2004;**38**(18): 3980-3992

[16] Simeonov V, Stratis J, Samara C, Zachariadis G, Voutsa D, Anthemidis A, et al. Assessment of the surface water quality in northern Greece. Water Research. 2003;**37**(17):4119-4124

[17] Boyacioglu H. Spatial differentiation of water quality between reservoirs under anthropogenic and natural factors based on statistical approach. Archives of Environmental Protection. 2014; **40**(1):41-50

[18] Varol M, Gökot B, Bekleyen A, Şen B. Water quality assessment and apportionment of pollution sources of Tigris River (Turkey) using multivariate

statistical techniques—A case study. River research and applications. 2012; **28**(9):1428-1438

[19] Ouyang Y. Evaluation of river water quality monitoring stations by principal component analysis. Water Research. 2005;**39**(12):2621-2635

[20] Bengraïne K, Marhaba TF. Using principal component analysis to monitor spatial and temporal changes in water quality. Journal of Hazardous Materials. 2003;**100**(1–3):179-195

[21] Kazi TG, Arain MB, Jamali MK, Jalbani N, Afridi HI, Sarfraz RA, et al. Assessment of water quality of polluted lake using multivariate statistical techniques: A case study. Ecotoxicology and Environmental Safety. 2009;**72**(2): 301-309

[22] Juntunen P, Liukkonen M, Lehtola M, Hiltunen Y. Cluster analysis by self-organizing maps: An application to the modelling of water quality in a treatment process. Applied Soft Computing. 2013;**13**(7):3191-3196

[23] Jolliffe I. Principal component analysis: Wiley online. Library. 2005

[24] Jackson JE. A user's Guide to Principal Components: John Wiley & Sons; 2005.

[25] Lawless HT, Data Relationships HH. Multivariate Applications. Springer; 2010

[26] Kim J-O, Mueller CW. Introduction to Factor Analysis: What it Is and How to do it. SAGE Publications; 1978

[27] Cadima J, Jolliffe IT. Loading and correlations in the interpretation of principle compenents. Journal of Applied Statistics. 1995;**22**(2):203-214

[28] Statistics DJC. Data Analysis in Geology. John Wiley & Sons; 1986

[29] Razmkhah H, Abrishamchi A, Torkian A. Evaluation of spatial and temporal variation in water quality by pattern recognition techniques: A case study on Jajrood River (Tehran, Iran). Journal of Environmental Management. 2010;**91**(4):852-860

[30] Mazlum N, Ozer A, Mazlum S. Interpretation of water quality data by principal components analysis. Turkish Journal of Engineering & Environmental Sciences/Turk Muhendislik ve Cevre Bilimleri Dergisi. 1999;**23**(1):19-26

[31] Haag I, Westrich B. Processes governing river water quality identified by principal component analysis. Hydrological Processes. 2002;**16**(16): 3113-3130

[32] Rangeti I. Determinants of key drivers for potable water treatment cost in uMngeni basin. Durban: Durban Universty of. Technology. 2015

[33] Panda UC, Sundaray SK, Rath P, Nayak BB, Bhatta D. Application of factor and cluster analysis for characterization of river and estuarine water systems – A case study: Mahanadi River (India). Journal of Hydrology. 2006;**331**(3–4):434-445

[34] Ge J, Ran G, Miao W, Cao H, Wu S, cheng L. Water quality assessment of Gufu River in three gorges reservoir (China) using multivariable statistical methods. Advance Journal of Food Science and Technology. 2013;**5**(7): 908-920

[35] Ragno G, Luca MD, Ioele G. An application of cluster analysis and multivariate classification methods to spring water monitoring Data. Microchemical Journal. 2007;**87**:119-127

[36] Johnson RA, Wichern DW. Education P. Prentice hall Englewood Cliffs, NJ: Applied multivariate statistical analysis; 1992

[37] Otto M. Multivariate Methods. Analytical chemistry Weinheim: Wiley-VCH; 1998

[38] Shrestha S, Kazama F. Assessment of surface water quality using multivariate statistical techniques: A case study of the Fuji river basin, Japan. Environmental Modelling & Software. 2007;**22**(4):464-475

[39] Basson MS, Rossow JD. Mvoti to Umzimkulu Water Management Area Overview of Water Resource Availability and Utilisation. Department of Water Affairs (DWAF); 2003. Report No.: PWMA 11/000/00/0203 Contract No.: PWMA 11/000/00/0203.

[40] Wilson AJ. Mvoti to Mzimkulu water management area situational assessment South Africa: A.J. Wilson & Associates International cc. In: cc AJWAI. 2001

[41] Reghunath R, Murthy TS, Raghavan B. The utility of multivariate statistical techniques in hydrogeochemical studies: An example from Karnataka. India. Water research. 2002;**36**(10):2437-2442

[42] Williams B, Brown T, Onsman A. Exploratory factor analysis: A five-step guide for novices. Australasian Journal of Paramedicine. 2012;**8**(3):1

[43] Kaiser HF. An index of factorial simplicity. Psychometrika. 1974;**39**(1): 31-36

[44] Parinet B, Lhote A, Legube B. Principal component analysis: An appropriate tool for water quality evaluation and management—Application to a tropical lake system. Ecological Modelling. 2004;**178**(3–4): 295-311

[45] Jianqin M, Jingjing G, Xiaojie L. Water quality evaluation model based on principal component analysis and information entropy: Application in Jinshui River. Journal of resources and ecology. 2010;**1**(3):249-252

[46] Ryberg KR. Continuous Water-Quality Monitoring and Regression Analysis to Estimate Constituent Concentrations and Loads in the Red River of the North, Fargo, North Dakota, 2003–05. 2006.

[47] Ryberg KR. Cluster Analysis of Water-Quality Data for Lake Sakakawea, Audubon Lake, and McClusky Canal, Central North Dakota, 1990–2003. 2007.

[48] Schertz TL, Alexander RB, Ohe DJ. The Computer Program Estimate Trend (Estrend), a System for the Detection of Trends in Water-Quality Data. U.S. Geological Survey, Water-Resources Investigations: Reston, Virginia; 1991

[49] Lettenmaier DP, Conquest LL, Hughes JP. Routine streams and rivers water quality trend monitoring review. Washington: University of Washington; 1982. Contract No.: Technical Report 75.

[50] Mateo-Sagasta J, Zadeh SM, Turral H, Burke J. Water Pollution from Agriculture: A Global Review. Executive Summary: Rome, Italy: FAO Colombo. Sri Lanka: International Water Management ...; 2017

[51] Dzwairo B. Modelling raw water quality variability in order to predict cost of water treatment [DTech thesis]. Pretoria: Tshwane university of. Technology. 2011

[52] Walmsely RD, Butty M. Limnology of some selected south African impoundments. Pretoria: Water Research Commission/National Institute for water research Council for Scientific and. Industrial Research. 1980

[53] Neysmith J. Investigating Non-regulatory Barriers and Incentives to Stakeholder Participation in Reducing

Water Pollution in Pietermaritzburg's Baynespruit [Masters Degree]. Pietermaritzburg: University of KwaZulu-Natal; 2008

[54] Geographic N. National Geographic. línea] Available: http://www nationalgeographic com es/ historia/grandes-reportajes/los-barcos-de-losfaraones_8270/3[Último acceso: 20 Enero 2017]. 2017.

Chapter 6

Riparian-Buffer Loss and Pesticide Incidence in Freshwater Matrices of Ikpoba River (Nigeria): Policy Recommendations for the Protection of Tropical River Basins

Azubuike Victor Chukwuka and Ozekeke Ogbeide

Abstract

The unregulated use of watersheds for agriculture negatively impacts the quality of river basins. In particular, the reduced quality of surface-waters, have been attributed to absence or poorly-decided riparian-buffer specifications in environmental laws. To demonstrate suitable buffer-width for protection of surface water, sediment and benthic fish populations, five riparian areas with different vegetation richness and buffer-width were selected within an organochlorine pesticide (OCP)-impacted watershed using the Normalized Differential Vegetation Index (NDVI) and multiple buffer analysis respectively. Mean OCP levels in surface water, sediment and fish sampled at each riparian stations showed site-specific differences with markedly higher levels of α-BHC, β-BHC, δ-BHC, p,p′-DDD and total pesticide residues at stations with least riparian cover. The principal component analysis further revealed more OCPs associating with sediment and fish from stations having smaller buffer-width and sparse riparian vegetation. Stations with wider buffer-width of at least 120 m provided greater protection to adjacent surface water and benthic fish populations. While this study recommends riparian buffer-widths for a typical tropical environment, further research which assesses other contaminant types in aquatic matrices adjacent to different riparian environments would be valuable and informative for regulatory guidance and strategic protection of ecosystem services.

Keywords: riparian reserve, labile pesticides, watershed, environmental policy, unsupervised imagery classification, NDVI

1. Introduction

The ecological integrity of watersheds is critical for normal ecosystem functioning of associated river systems [1, 2]. Riparian areas of fluvial systems consist of natural buffer vegetation which spans both sides of the stream bank and function in mitigating the impact of flood, sedimentation, and nutrient run-off from adjoining cultivated land areas [3, 4]. However, the dynamics of rapid urbanization, lack of environmental education among the citizenry, and outright government negligence which have put such environments under anthropogenic pressure is an issue of growing concern [5, 6].

IntechOpen

River basins are typically exposed to anthropogenic pressure via agricultural land-use in which attempting to harness fertile areas of alluvial deposits around fluvial systems, have accelerated the depletion of riparian vegetations [2, 7]. Such disruption of riparian landscape increases access of contaminants of anthropogenic origin into adjacent surface water, thus putting biodiversity and human populations at risk [8]. The use of organochlorine pesticides as a choice chemical for agricultural applications in tropical developing countries despite its ban still constitutes a concern due to its persistence and toxicity to non-target species [9, 10]. In agricultural catchments, the fate of pesticides including the uptake, accumulation and persistence within environmental matrices depends on its hydrophobicity or n-octanol–water partition coefficient (log K_{ow}) [11]. Several studies have demonstrated the ability of wide riparian vegetation areas to buffer adjacent surface water by filtering pesticide-laden runoff and top-soil originating from agricultural catchments [12, 13], with small buffer widths conferring little protection [13, 14].

Recent advances in spatial analysis and fluvial remote sensing, allow better assessment of such extraction of information regarding river basins behavior under natural and human-induced disruptions on a larger scale compared to *in situ* assessments alone [15]. Data ranging from low, medium and high-resolution imagery obtained from Earth Observation (EO) satellites [16, 17], have enabled the complementary use of indices to efficiently identify degraded riparian areas and allow prioritization of river basins for strategic protection [18–20]. In developing climes, the effective management of riparian ecosystems has been limited by weak, poorly constituted or outright non-existent policies, which have aggravated issues of riparian exploitation and its imminent loss [21, 22]. In Nigeria, the National Environmental (Watershed, Hilly, Mountainous and Catchment Areas) Regulations, 2009, is the major watershed-specific regulatory instrument used by the National Environment Standards and Regulations Enforcement Agency (NESREA), to monitor and enforce national laws and domesticate international agreements on the watershed [21]. Although this regulatory document describes watershed as 'the total land area that drains directly or indirectly into a particular stream or river', it makes no mention of key parameters like riparian zone or buffer width. Such documents exemplify regulatory instruments with limited scientific scope and coverage of their subjects' focus. Existence of such laws in developing countries has been attributed to limited research information and empirical data necessary for developing context-specific guidelines for such environments [5]. Similar cases of laws and regulations with forest and watershed specifications not backed by empirical studies or proof of its suitability for the environment concerned have also been documented [6, 23].

Anthropogenic pressures which accelerate the degradation of river basins are particularly severe in African urban environments [8, 24]. A critical feature of river basin resilience is its lateral connectivity between the river channel and its watershed [25]. The connecting stream network and volume of lateral connectivity (stream orders) within a river basin landscape, is a reliable proxy for discerning watershed disruption [26] and is greatly reduced via agricultural activities including soil leveling for irrigation [27]. Loss of riparian areas and altered pollution dynamics within watersheds due to agriculture and the use of pesticides is also typical of sub-Saharan areas, including Nigeria [9]. Evidence of recent use of banned pesticides, including OCPs, with environmental concentrations that portend risks to both humans and aquatic wildlife, has spotlighted the Ikpoba agricultural catchment within Edo area [28, 29]. This study explores the complementary use of medium resolution (Landsat 8, OLI) and high resolution (Google earth) imageries to describe riparian loss within the Ikpoba watershed. Besides, we also sought to demonstrate effective riparian buffer-width by relating pesticide incidence in adjacent surface water, sediment and benthic fish to buffer-width within the agrarian catchment of Ikpoba. We hypothesize that an effective buffer-width and density of riparian cover within the Ikpoba watershed will limit the incidence of pesticides in adjacent surface water, sediment and fish.

2. Materials and methods

2.1 Study area

Ikpoba River is a sixth-order river located between Lat. 6° 19′ 12“ N, long 5° 24’ 0”E and Lat. 6° 22′ 48“N, long 5° 51’ 7.2”E in Benin City, Edo State, Nigeria (**Figure 1**) [30]. The Ikpoba catchment area which extends laterally along the rivers longitudinal continuum is a notable urban watershed within the heart of Benin city, Edo state. Ikpoba's river riparian area offers local communities with diverse resources including fisheries and domestic water supply. The study section has a width of 10 to 12 meters' watercourse and characterized by a predominance of indigenous plant species i.e. bamboo trees (*Bambusa vulgaris*) that make up the riparian vegetation [31]. However, most of this vegetation has been lost due to massive deforestation for agricultural practices and other anthropogenic activities. The favorability of the landscape area for agricultural purposes could be attributed to its rich alluvial landscape.

2.2 Geospatial analysis

To demonstrate the role of riparian vegetation in preserving the quality of adjacent surface water, a prior spatial description of the area and determination of riparian areas that exemplify the different degree of riparian alteration was required. To achieve this, the geospatial analysis workflow is detailed sequentially by sections. The schematic representation of the workflow is given in Appendix I.

2.2.1 Data acquisition

Viable spatial patterns of land-use and land cover of the Ikpoba riparian area were first visualized within the google earth database before acquisition using the snapshot tool of Google earth 7.1 software. The images offered by the software originate from both satellites and aerial photography with a repeatability update ranging from 6 months to 5 years [32]. This repetitive frequency of updating images over time makes them an effective and reliable alternative to non-updated hard-copies of maps and surveys available at the local and state government repositories.

The watershed and riparian area of Ikpoba river were mapped at a fine spatial scale from Google Earth Imagery with discernable land use and landcover features particularly the river course and aquatic vegetation within the watershed. Confirmatory ground-truthing was carried out using expert knowledge of the area to ascertain the general layout, and interspersed anthropogenic activities including farming activities and built-up areas (**Figure 1a**). Detailed photographs of the area were also taken as evidence of riparian features (Appendix II).

2.2.2 Land-use land cover classification

To improve the spatial visualization of the watershed and riparian extent of Ikpoba river, and highlight the finer details of land use/cover features-classes, high spatial resolution raster image acquired from Google Earth was subjected to Iso

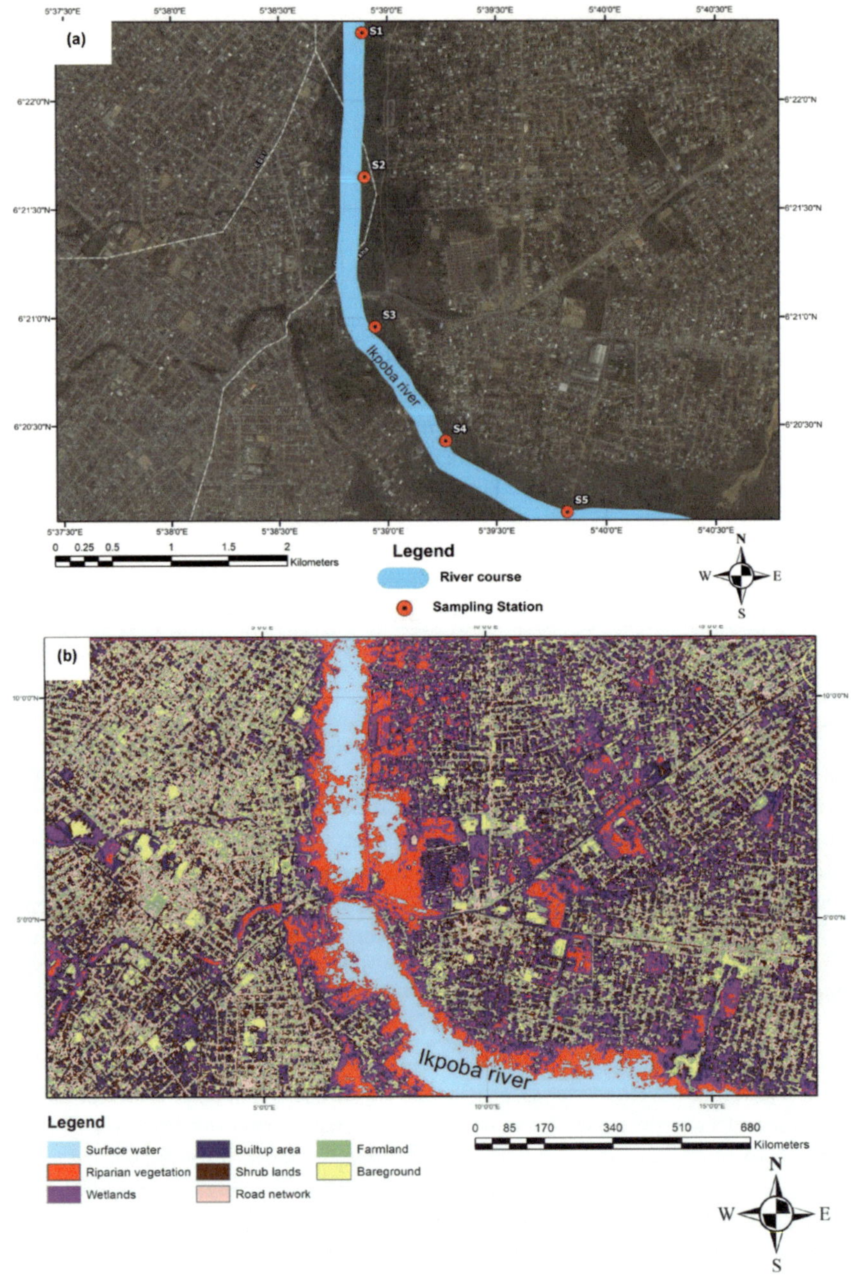

Figure 1.
Map of Ikpoba River, showing (a) sampling stations and (b) land-use land-cover classes.

cluster unsupervised classification in ArcMap 10.4 [33]. The choice of google earth imageries over traditional Landsat imageries for land-use land-cover classification is attributable to its relatively lesser cost of acquisition [34], better discriminating capacity of google earth imageries compared to Landsat and parallel classification testing which rates the classification accuracy of google earth imagery to be within range attainable by other high-resolution imagery like QuickBird [35]. The output from the unsupervised classification highlighted the riparian extent of the Ikpoba area and revealed eight land-use land-cover features i.e. surface water, riparian vegetation, wetlands, built-up area, shrublands, road-network, farmland and bare-ground (**Figure 1b**).

2.2.3 Vegetation analysis

To quantify the vegetation richness of the riparian area and establish an empirical basis for the difference in riparian vegetation along the Ikpoba river, the Normalized Difference Vegetation Index (NDVI) was calculated. For this purpose, Landsat 8 imagery was acquired from the United States Geological Survey (USGS) website. The index was calculated by imputing the red band (R) and near-infrared band (PIR) contained therein into the raster calculator of ArcGIS version 10.4., using Eq. 1 below.

$$\mathrm{NDVI} = (\mathrm{B4} - \mathrm{B3})/(\mathrm{B4} + \mathrm{B3}) \tag{1}$$

Where B3 and B4 is the red band and near-infrared band respectively.

This standardized vegetation index is sensitive to plant vigor and abundance by illustrating the disparity between the visible red band and the near-infrared band. The output of this analysis highlighted the rich and depleted riparian areas within the watershed, where green, yellow and red pixels extending laterally to the fluvial course was considered to be gradients of richness or depleted-ness of riparian vegetation (**Figure 2**) [36].

2.2.4 Creating stream orders

In addition to measures of riparian disruption and watershed dynamics, the stream order network for the Ikpoba area was assessed by processing digital

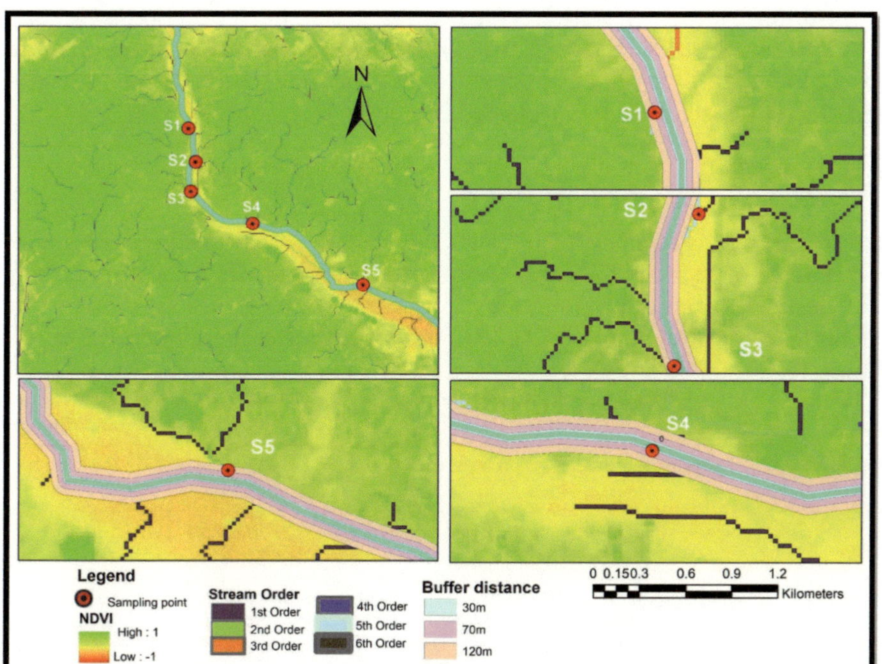

Figure 2.
NDVI analysis of riparian vegetation along Ikpoba river where station 1 (dense riparian vegetation on both sides of the river+ higher-order tributary), 2 (sparse riparian vegetation on one side of the river+2 lower-order tributaries), 3 (sparse vegetation buffer on one side of the river + one lower-order tributary), station 4 (moderate vegetation buffer on one side of the river+3 lower-order tributaries), station 5 (sparse vegetation buffer on both sides of the river+3 lower-order tributaries).

elevation models (DEMs) of the Ikpoba area in ArcMap® version 10.4 using ArcGIS Hydrology tools (Appendix I) [37, 38]. The DEMs of the Ikpoba area were also downloaded from the United States Geological Survey (USGS) website. The classification of stream order offered a standardized approach for defining the volume/size of the river (i.e. reaching length x width x depth under average flow conditions) and the potential hydrological ability of each stream within the catchment. Using the "Strahler" method available under hydrological tools in the ESRI ArcGIS 10.4 software, [39, 40] very small streams were categorized as a stream order of "1–3", while larger rivers were designated a stream order of "4–6". Stream order used as a measure of landscape connectivity and watershed disruption highlighted large incidences of lower-order streams entering the river system indicating a disconnectivity from headwaters and vulnerability to pollutants from overland runoff [8, 26]. Incidence of higher-order rivers emptying into the Ikpoba river (6th order stream) was taken as indices of relative connectivity to headwaters from the watershed and lesser vulnerability to overland runoff and pollutant-laden discharge.

2.2.5 Riparian station delineation

Using the color gradient scale of the NDVI analysis to establish richness and sparseness of riparian vegetation along Ikpoba river, control station was assigned to an area with ecologically desirable features i.e. dense riparian vegetation on both sides of the river+ higher-order tributary (station 1). Station 2 was assigned to an area of the watershed having sparse riparian vegetation on one side of the river+2 lower-order tributaries. Station 3 was assigned to an area characterized by sparse vegetation buffer on one of the rivers + one lower-order tributary. Station 4 was located within an area having moderate vegetation richness on one side of the river +3 lower-order tributaries, station 5 (sparse vegetation buffer on both sides of the river+3 lower-order tributaries) (**Table 1**). In total, five different predetermined stations/areas of the watershed approximately 1 km apart, with different degree of altered riparian vegetation were selected.

2.2.6 Buffer analysis

For this study, riparian buffer [41] is "a strip or area of vegetation adjacent to a river or stream of sufficient width necessary to remove nutrients, sediment, organic matter, pesticides, and other pollutants from surface runoff and subsurface flow by deposition, absorption, plant uptake, and other processes, thereby reducing

Sampling station	Riparian Feature description	Coordinates	Distance to the next sampling area
1 (control site)	Rich riparian vegetation on both sides of the river	6°22′18.76"N 5°38′53.10″E	1.2 km
2	sparse riparian vegetation on one side of the river+2 lower-order tributaries	6°21′38.81"N 5°38′53.62″E	1.29 km
3	sparse vegetation buffer on one of the rivers + one lower-order tributary	6°20′57.38"N 5°38′56.46″E	1.13 km
4	moderate vegetation buffer on one side of the river+3 lower-order tributaries	6°20′25.70"N 5°39′15.88″E	1.19 km
5	sparse vegetation buffer on both sides of the river+3 lower-order tributaries	6°20′5.83"N 5°39′49.27″E	

Table 1.
Sample site designation for different riparian features along the Ikpoba watershed.

pollution and protecting surface water and subsurface water quality, which are also intended to provide shade to reduce water temperature for improved habitat for aquatic organisms and supply large woody debris for aquatic organisms and habitat for wildlife".

To determine the width of riparian buffer at each station, and to allow comparison with the minimum recommended riparian widths specified for the protection of water quality and improvement of in-stream biodiversity i.e. 30 m, 70 m and 120 m [4], a multiple buffer ring features was created around the stream feature using the 'multiple buffer' command in ArcGIS version 10.4.

To extract total vegetation richness around each riparian station, NDVI raster was clipped for each station within a 120 m buffer radius. Mean pixel threshold values for each area within the 120 m buffer zone was extracted from the attribute table of the clipped NDVI raster image. The difference in mean vegetation richness of the riparian stations was confirmed using mean pixel thresholds extracted for each station within the 120 m buffer radius and subjected to one-way ANOVA analysis (Appendix III).

Table 1: Sample station designation for different riparian features along the Ikpoba watershed.

2.3 Sample collection and analysis

2.3.1 Water, sediment and biota samples collection

Every fourth night, from January 2016 to June 2016, samples of water (n = 60, 12 samples per station) and sediment (n = 60, 12 samples per station) were obtained from the five predetermined points along the river. As earlier specified, the predetermined points were runoff generating points long the watershed having different riparian features. Samples were gathered using an Eckman grab from three riverbeds at depths of 5 m for sediment and a hydro-bios sampler at depths of 0.3 m in a pre-cleaned 1 L glass bottle for water samples [42]. Impurities in water samples were eradicated using fiber-glass filters. Five (5) samples of *Clarias gariepinus,* were collected monthly from each riparian station. A total of 150 fishes (n = 30 per station) were collected for 6 months, January –June 2016. The Catfish, *Clarias gariepinus,* was caught using gillnets with 50–55 mm meshes between opposite knots. All samples were properly labeled and placed in an ice-chest at 4 °C afterwards transported to the laboratory for further analysis.

2.3.2 Extraction of samples

A method described by Hladik and McWayne [43] and Steinwandter [44] was used for the extraction of OCP residues in sediment samples and fish samples respectively. In summary, 15 g of the homogenized sediment samples were obtained by air-drying and sieving through a 2-mm sieve, while 25 g of edible portions of *Clarias gariepinus,* with 100 ml of acetone was homogenized for 20 min at 100 rpm. Homogenized samples were mixed with dichloromethane (DCM) and n-hexane (10:90) and sonicated for about 20–25 min at 50 °C. Similarly, OCP residues in water samples were extracted using the same solvent mix (n-hexane and dichloromethane (DCM) (50:50)), based on a method vividly described by USEPA [45]. Here, a volume of 250 ml water samples was mixed with 60 ml of the prepared solvent mix. The resulting solution was filtered with a separator funnel and then concentrated to 10 ml in a water bath at 35–40 ° C with the aid of a rotary evaporator. Clean up of water, sediment and fish extracts was achieved using a florisil solid-phase extraction (SPE) method [45]. This florisil column was equipped with 15% hexane as an eluting mixture Elutes were then transferred in vials to gas chromatography and later concentrated into a final volume of 1 mL using a rotary evaporator operating at six mbar at 30 °C.

2.3.3 Analysis of samples

All extracts were analyzed for OCPs using the EPA 8081 pesticide standard mix containing the following pesticides (aldrin, endrin, dieldrin, (α, β, γ, δ), endosulfan (I, II & sulfate), chlordane (α and γ-chlordane), isomers of DDT (p,p′-DDD, p,p′-DDE, p,p′-DDT), endrin aldehyde and heptachlor and heptachlor epoxide. Choice of pesticides analyzed was informed from a previous study which revealed predominant use of OCPs within the same catchment area [28]. Concentrations of these OCPs were determined using Gas chromatography (Hewlett-Packard (hp) 5890 Series II) equipped with a 63 Ni electron capture detector (ECD). Conditions of this instrument were as follows: chromatographic separation was obtained using a VF-5 ms capillary column of30 mm x 0.25 mm internal diameter x 0.25 μm film thicknesses with a 1 m retention gap (0.53 mm, deactivated). The carrier gas for this instrument was helium and it was used to run 1.5 lL aliquot of the sample with a splitless injection mode. At standard pressure, the flow rate was kept constant at 29 ml/min while the temperature of 250 °C. The initial oven temperature was set at 60 °C for 2 min and ramped at 25 °C /min to 300 °C for 5 min and allowed to stay for 15 min giving a total run time of 58 min. The detector temperature was 320 °C and was held for 5 min.

2.3.4 Quality assurance

To ensure the reliability and precision of the tests, all analyzes are subject to strict quality control, quality assurance and precautionary procedures. All materials used for the collection of water samples was pre-cleaned with ethanol. The research used double-distilled deionized water and analysis grade reagents while glassware was properly purified before the experiment. Instruments were calibrated using blank reagents determinations. The validity, as well as reliability of extraction and cleanup methods, was ensured by the recovery of an internal standard. For this research, analytical grade decachlorobiphenyl was purchased from Sigma Aldrich and used as an internal standard. Overall, with 10 ng/g level of fortification, the pesticide recovery rate was 85–90% for all matrices, with method detection limits (MDL) of 0.005 μg/g/dw, 0.01 μg/g/dw and 0.01 μg/l for sediment, fish and water respectively.

2.4 Statistical analysis

Statistical analysis was carried out on the data obtained from across each station. Analyses were performed by the use of Statistica® 12.0 (Stat Soft Inc. USA) and Microsoft Excel 2013 for Windows. One-way ANOVA test compared values and significant differences in mean were determined using Duncan's test ($p < 0.05$). Following the large number of inter-related pesticide variables in fish, sediment and surface water, Principal Components Analysis (PCA) was used to organize variables into sets of rotated inter-correlated variables (principal components or PCs) on the basis of a correlation matrix. The interrelated output of variables from the PCA are also referred to as linear combinations of the original variables that form the axes, used to construct a biplot. Each of the PCs or axes account for the variances in data, and only axes accounting for large variances (mostly PC 1 and PC 2) were considered for interpretation [46]. Biplots were used to visualize the association of pesticides in fish, sediment or water with particular riparian areas. Close proximity of pesticides species with particular riparian stations within the ordination space of the biplot was termed strong association and was indicative of high concentrations of that pesticide in that riparian area [47].

3. Results

3.1 Pesticide concentration in environmental matrices across riparian stations

Results of pesticides residues in fish samples obtained from Ikpoba river shows that station 2 had the highest concentration of total pesticide residues. There was no significant difference ($p \leq 0.05$) in total pesticide concentration in fish samples from all stations. δ-BHC (0.00082 ± 0.00024 μg/kg), α-BHC (0.00608 ± 0.00108 μg/kg), p,p′-DDD (0.00150 ± 0.00027 μg/kg), endosulfan-1 (0.00084 ± 0.00032 μg/kg), and β-BHC (0.00236 ± 0.00052 μg/kg) were observed to have the highest concentrations in stations 1,2,3,4, and 5 respectively. Pesticide levels in fish across sampling stations of riparian stations showed marked levels of α-BHC and β-BHC in fishes from station 2. Again, total pesticide concentration was higher at station 2, followed by station 5 and station 3. i.e. station 2 > station 5 > station 3 > station 4 > station 1 (**Figure 3a,b**).

In sediment samples, it was observed that concentrations of p,p′-DDT (0.00155 ± 0.00026 μg/kg), BHC (0.00284 ± 0.00048 μg/kg), heptachlor (0.0036 ± 0.00061 μg/kg), dieldrin (0.0022 ± 0.00038 μg/kg) and p,p′-DDD (0.0033 ± 0.00055 μg/kg) were highest for stations 1, 2, 3, 4 and 5 respectively. Overall Station 2 and 3 had higher concentrations of total pesticide residues compared with the other stations although there was no significant difference in total pesticide residues between the five stations. In sediment, total pesticide concentration was markedly high at station 2 and station 3, with lowest values at station 4 i.e. station 2 = station 3 > station 5 > station 1 > station 4 (**Figure 3c,d**).

For water samples, the highest concentration of total pesticide residues was observed at station 2. At station 1 (0.00041 ± 0.00006 μg/L) and 2 (0.00086 ± 0.00014 μg/L), α-BHC had the concentration of pesticides while δ-BHC and β-BHC were highest at stations 4 (0.00024 ± 0.00006 μg/L) and

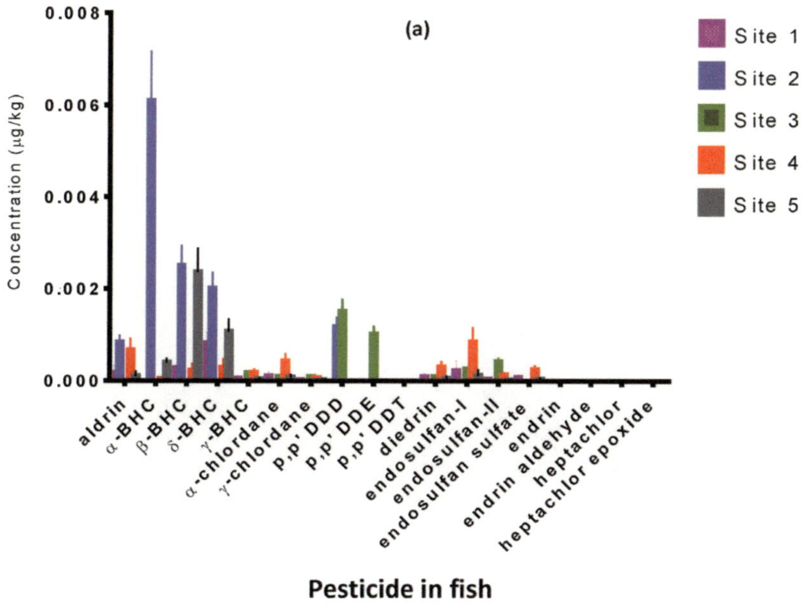

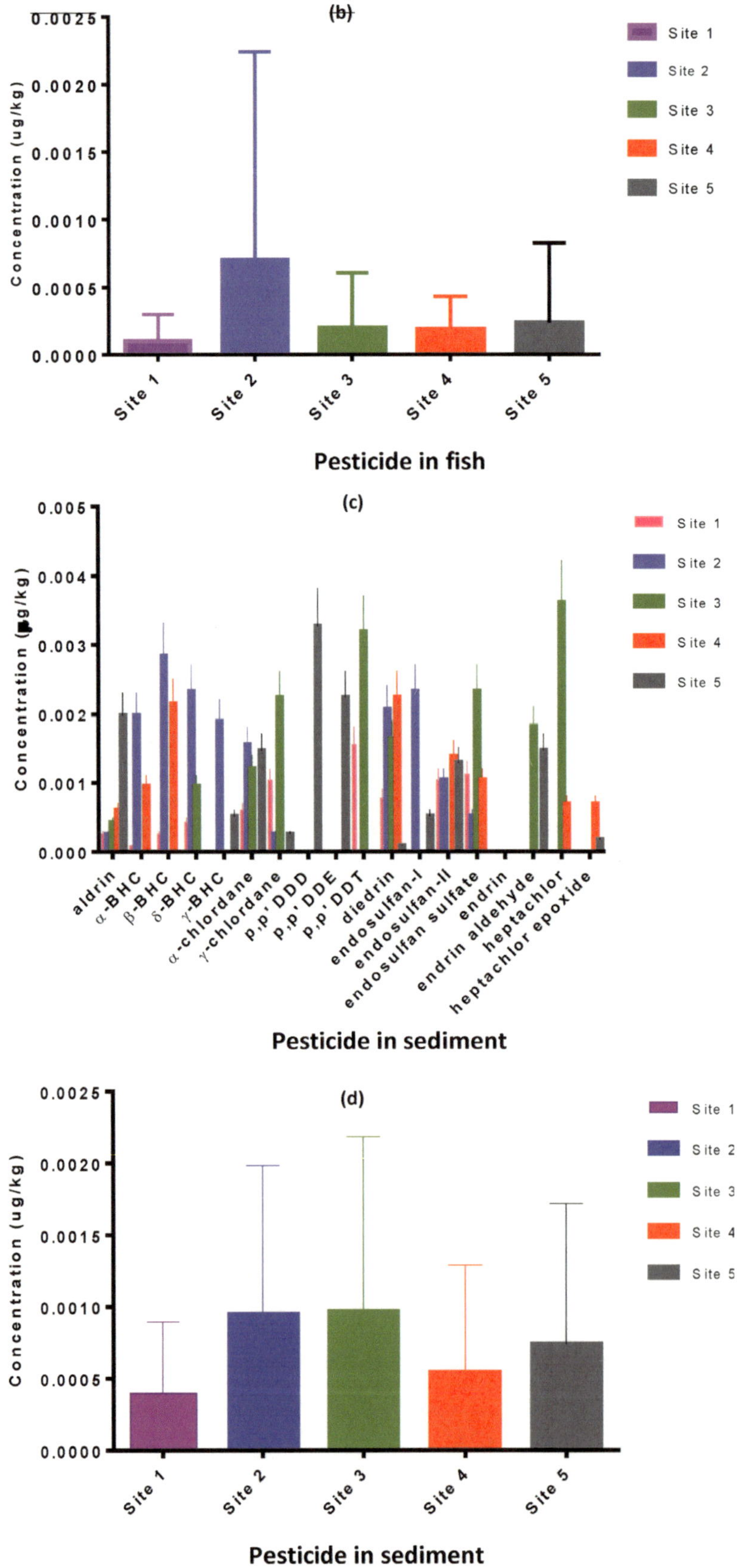
(b)
Concentration (ug/kg)
0.0025
0.0020
0.0015
0.0010
0.0005
0.0000
Site 1
Site 2
Site 3
Site 4
Site 5
Pesticide in fish
(c)
Concentration (µg/kg)
0.005
0.004
0.003
0.002
0.001
0.000
aldrin
α-BHC
β-BHC
δ-BHC
γ-BHC
α-chlordane
γ-chlordane
p,p' DDD
p,p' DDE
p,p' DDT
diedrin
endosulfan-I
endosulfan-II
endosulfan sulfate
endrin
endrin aldehyde
heptachlor
heptachlor epoxide
Site 1
Site 2
Site 3
Site 4
Site 5
Pesticide in sediment
(d)
Concentration (ug/kg)
0.0025
0.0020
0.0015
0.0010
0.0005
0.0000
Site 1
Site 2
Site 3
Site 4
Site 5
Pesticide in sediment

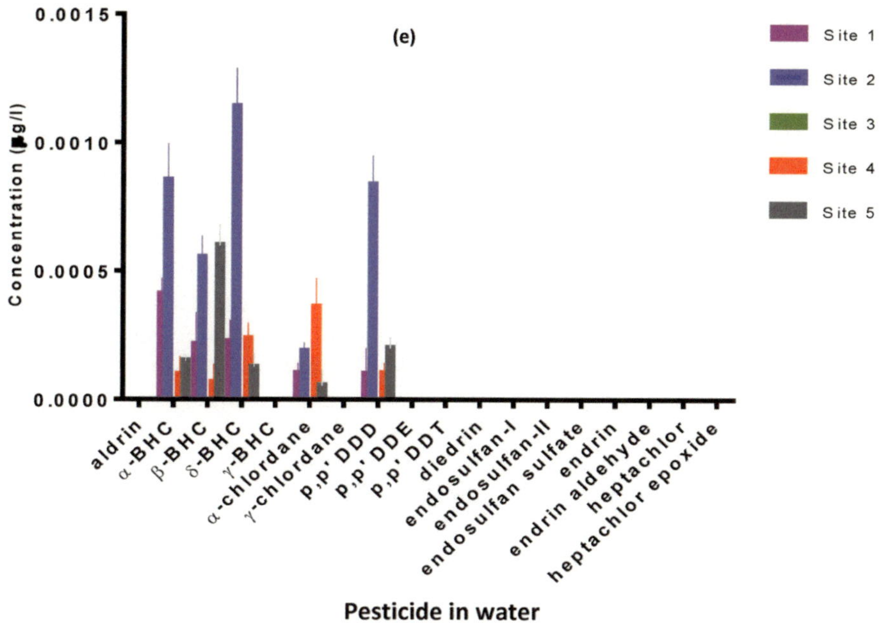

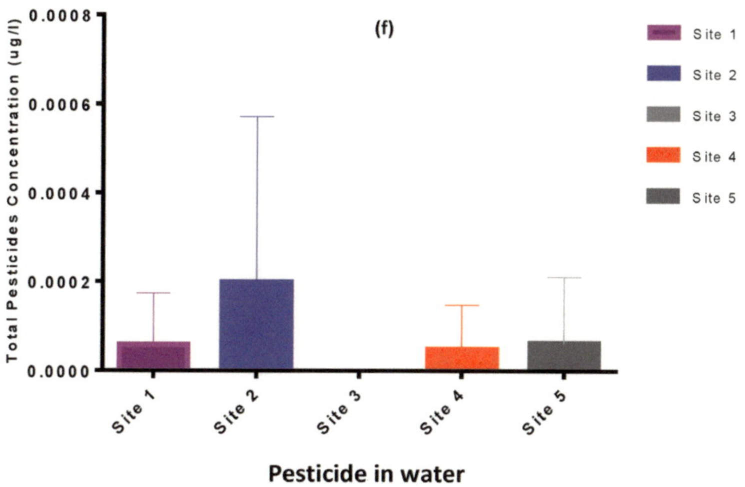

Figure 3.
Single pesticides and total pesticide concentrations respectively for (a-b) fish (c-d) sediment (e-f) water across riparian sites.

5 (0.00060 ± 0.00008 μg/L). Pesticide concentration in surface water across riparian stations in this study, revealed markedly high levels of α-BHC, β-BHC, δ-BHC, and p,p′-DDD at station 2, while other stations showed markedly lower values (**Figure 3e,f**).

3.2 Multivariate relationship between riparian profile and pesticide incidence

Multivariate tests using principal component analysis (PCA), recorded varying patterns of associations between riparian stations and incidence of pesticides in sediment, surface water and fish across riparian stations. Describing occurrence of pesticides according to n-octanol–water partition coefficient (log K_{ow}) i.e. hydrophilic (>4.5) or hydrophobic (<4.5) (Appendix IV), gave better insight into pesticide mobility and occurrence within environmental matrices of the Ikpoba catchment.

For sediment, station 2 showed strong positive associations with the greatest number of pesticide types (α-BHC, β-BHC, γ-BHC, Endosulfan 1, δ-BHC and dieldrin). Five of them were hydrophilic pesticides while dieldrin was the only hydrophobic pesticide (5:1). Contrastingly, Station 3 showed a strong positive association with a greater number of hydrophobic pesticides (3:1), while endosulfan was the only hydrophilic pesticide detected in this area. Station 4 also showed a positive association with a greater number of hydrophobic (2:1), with endosulfan II as the only hydrophilic pesticide strongly associated with sediment from this area. All three pesticides positively associated with sediments at station 5 were all hydrophobic pesticides. While endrin aldehyde did not show any station-specific association or pattern of occurrence, sediments from station 1 did not show any marked association with any pesticide type (**Figure 4**).

For pesticide incidence in fish across riparian stations, the PCA revealed a strong positive association between station 2 and fish with the predominant occurrence of high labile pesticides (β-BHC, α-BHC, δ-BHC); station 4 showed positive associations with both low labile (dieldrin, a-chlordane and Aldrin) and high labile (endosulfan 1, endosulfan sulfate) pesticides. Fish samples from station 3 were only strongly associated with p,p′-DDE (a low labile pesticide). Station 1 and 5 showed no marked or weak association with any pesticide species (**Figure 5**).

PCA for pesticides in surface water across riparian stations revealed that surface water from station 2 was strongly associated with the greatest number of pesticides consisting of both high labile (α-BHC, δ-BHC) and low labile (p,p′-DDD) pesticides. Station 4 surface water samples showed a strong association with single pesticide a-chlordane. Surface water samples from station 1, 3 and 5 did not show any strong association with any of the pesticide species detected in this study, indicating a generally low level/incidence in surface water from these stations (**Figure 6**).

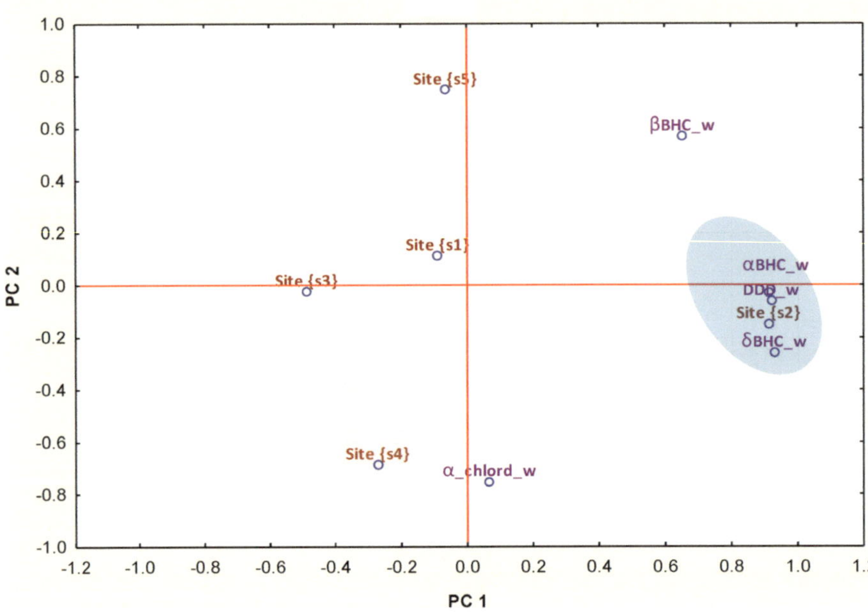

Figure 4.
PCA biplot for contaminants in sediment across sites along the Ikpoba watershed.

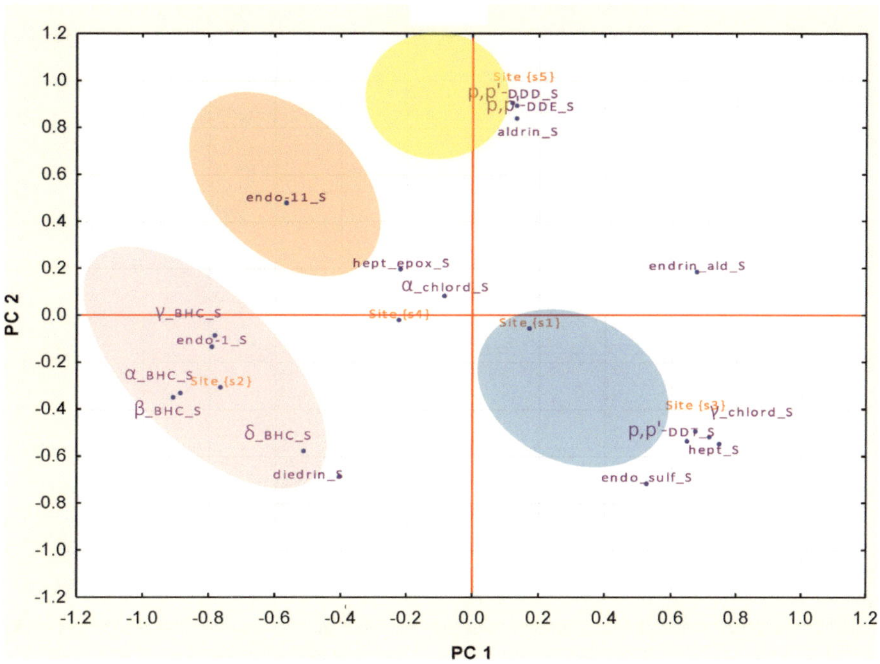

Figure 5.
PCA biplot for contaminants in fish across sites along the Ikpoba watershed.

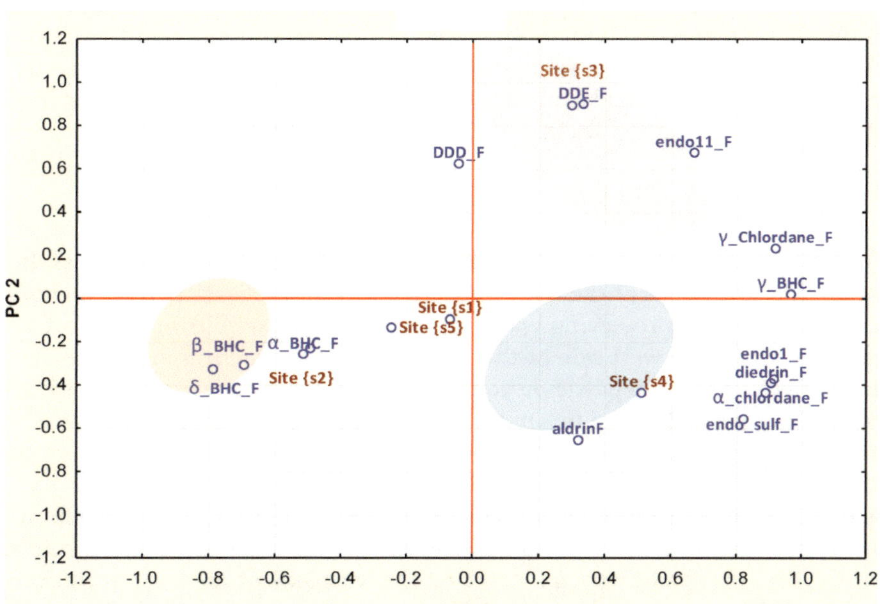

Figure 6.
PCA biplot for contaminants in surface water across sites along the Ikpoba watershed.

From the combined PCA plot for pesticide concentrations in fish, sediment and water, there was a notable incidence of low labile pesticides, γ-chlordane and p,p′-DDE in fish flesh sampled from station 3 that was positively associated with γ-chlordane and DDT in sediment (**Figure** 7).

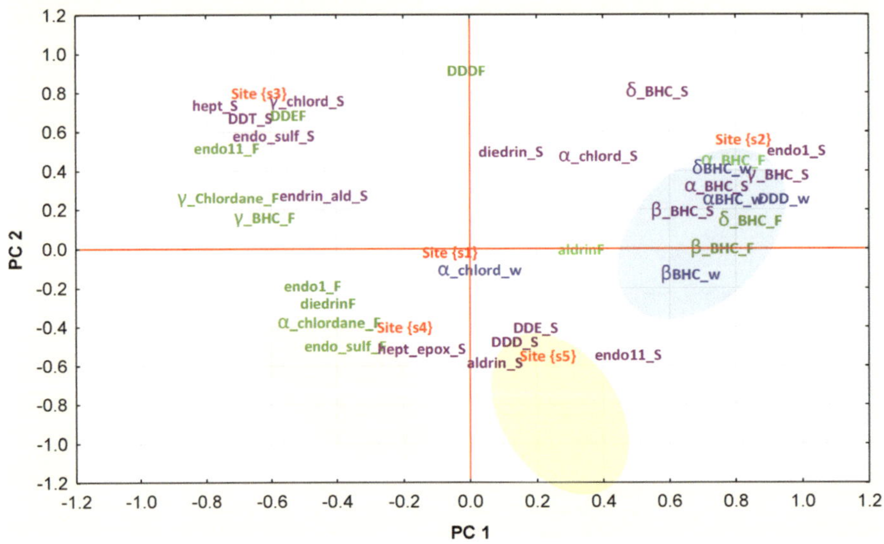

Figure 7.
PCA biplot for contaminants in water, sediment and fish combined across sites along the Ikpoba watershed.

4. Discussion

Riparian buffers are vital links between terrestrial and aquatic ecosystems and they regulate the flow of species, energy, and various materials including contaminants between these ecosystems [48]. Under normal conditions of sufficient buffer-width and vegetation density, riparian vegetation buffers have the potential to remove and detoxify pesticides in runoff [13], however, only a few studies have examined the fate of pesticides in riparian areas [49].

4.1 Pollutant incidence in sediment and surface water

The patterns of pesticide incidence in surface water and sediment across different riparian profile areas of the watershed give a first impression of the role of riparian vegetation in preserving water quality. Station 2 with over 120 m of sparse vegetation only on the west side of the river, showed the highest incidence of pesticide species and total pesticide concentration in surface water. Higher incidence of pesticides in surface water adjacent to agricultural catchments has been attributed to recent and probably ongoing pesticide applications within the area [11, 50].

Sparse riparian vegetation cover increases the likelihood of soil compaction/bank shearing and ultimately, increased runoff into surface water from the affected area [51], thus, the sparse riparian cover on the west side of the river at station 2 may be implicated in the high occurrence of pesticides at this station. The lower pesticide incidence in surface water recorded for other riparian stations with <120 m rich vegetation suggests the absence of recent pesticide applications within that axis of the watershed. Surface water adjacent to buffer strips with intact riparian vegetation has been found to have better water quality compared to stations adjacent to buffer strips with little or no vegetation [52].

The difference in OCP concentration and types in sediments sampled at the five riparian stations are also relatable to the narrow riparian buffer (< 120 m) and sparse vegetation cover. Higher magnitude of above-ground sediment flow and the erosion of streambanks has been associated with bare topsoil in watersheds covered by little or no vegetation [40]. This, in turn, amounts to the loss of valuable soil and acreage and the resultant loss in quality of adjacent surface water [53]. While efficient buffer widths have been shown to differ in application, wider buffers are considerably more useful for ecosystem protection than narrow ones. The 120 m buffer-width highlighted in this study for the Ikpoba watershed is consistent with some national guidelines with width recommendation of 10–100 m on each river-bank depending on hydrology and vulnerability of landscape [5].

4.2 Pollutant incidence in fish

From this study, the varied occurrence of OCPs in the flesh of the catfish (benthic fish) across riparian stations was relatable to the size of riparian-buffer widths. From the combined PCA plot, the high incidence of high labile OCPs (β-BHC, α-BHC, δ-BHC) in fish at station 2 was strongly associated with the same labile pesticide species in surface-water and sediment. This observation readily suggests that the smaller buffer-width at station 2, allowed greater OCP transport via runoff into surface water, and sediment, and eventual uptake by benthic fish. The co-occurrence of high labile OCPs in sediment and surface water portends increase risks to aquatic species because they are readily taken up by gills [11, 54]. Ecological risk assessment based on observed concentrations in water and sediment in Ikpoba river showed potential for risk to the different trophic levels (algae, daphnia and fish) inhabiting the river [28].

Also, from the combined PCA plot, the incidence of low labile pesticides, γ-chlordane and p,p′-DDE (a metabolite of DDT) in fish flesh sampled from station 3 was positively associated with γ-chlordane and DDT in sediment. The low labile (hydrophobic) characteristics of these OCPs readily suggests that incidence of these OCPs into the river sediment may have largely occurred via sediment erosion, and washing away of topsoil into adjacent regions of the river [11]. The coincidence of the same labile OCPs in sediment and flesh portends that uptake by benthic fish could be more likely to occur via benthic trophic interactions within its habitat range [54]. Such uptake in local fish fauna could culminate to reduced growth, altered reproduction and recruitment in local populations, and major shifts in community structure and health of local fauna [40, 55, 56]. Freshwater habitats in wet tropics support significant aquatic biodiversity [57], thus shrinking riparian vegetation would disrupt local assemblages of biodiversity, macro-invertebrates, and vertebrates that feed on them [58].

The clear discrimination between fish sampled from different stations based on OCPs in flesh, not only depicts that the predetermined areas of the watershed are indeed experiencing different pollutant traffic and transport but also establishes that the fish population from the different stations are not intermingling populations. This discrete uptake of OCPs depicts site-fidelity of species. It an ecological feature that prevents the additional physiological costs of exploring new areas [59], and is advantageous for increased individual survival and recruitment [60, 61] and better population viability [62]. The distances between the Ikpoba riparian stations (i.e. having a minimum of 1 km intervals) juxtaposed with the discrete OCP uptake by the catfish populations, and inferences of site-fidelity are consistent with studies that have proposed linear home range distance of approximately 1 km in similar benthopelagic fish [63]. The realization of site-fidelity for the catfish, a predominant benthic fish within the Ikpoba watershed portends that strategic conservation of different fish populations via designs of protected areas or implementations of fishing closures could be achieved if and when deemed necessary.

In general, the incidence of OCPs in fish, reveals that riparian stations with less than 120 m of buffer-width, cannot confer protection on biota in adjacent surface water. This finding demonstrated within a typical tropical catchment, is consistent with biodiversity studies in Latin America (tropical environment) and Southeast Asia that have recommended width thresholds of 40–200 m [5]. While our study corroborates other studies that attribute greater pesticide filtering capacity to wider buffer-width in riparian areas [13, 14], the use of NDVI, highlights vegetation vigor as a necessary feature to confer surface water protection. Many riparian studies have solely emphasized buffer width as a criterion for river protection [53, 55] while just a few have highlighted vegetation density or vigor [64, 65]. However, this study is a first report relating remotely sensed vegetation index to the filtering capacity and river protective features of riparian buffers in a typical tropical environment.

4.3 Stream order indices as a cofactor of riparian vulnerability

From this study, it was observed that riparian stations with the lowest occurrence of OCPs (station 1), not only had wide riparian buffer-width exceeding 120 m on both sides of the riverbank, and rich vegetation but received watershed drainage input from higher-order tributaries. Although, there was no linear relationship between the number of lower-order tributaries that characterized each riparian station, riparian stations with a greater incidence of OCPs in matrices received drainage from lower to intermediate-order tributaries. The association of lower-order drainage tributaries with narrow buffer-width areas and sparse vegetation presents a situation of altered hydrology due to disrupted watershed and loss of riparian vegetation. Such disrupted hydrology and tributary connectivity in agricultural catchment could have ecological implications, including limited dispersal corridors for biota and altered downstream delivery of substances within the watershed [66]. Hydraulic interactions during baseflows in watersheds of riparian areas allow the concentration of upstream flows, creating higher-order tributaries [67, 68]. As such, disrupted watersheds, allow predominance of lower-order or ephemeral tributaries which lack concentrated upstream flow and dominated by overland water flow [69]. As a result, these low-order tributaries are most prone to non-point sources of pollution, and increased pollutant load, and could significantly influence the quality of the receiving waters [70].

4.4 Riparian ecosystem services and policy advocacy

The occurrence of OCPs in sediments and fish adjacent to particular sampling stations indicates that the watershed has experienced a significant loss of riparian buffer-width and vegetation density resulting in what could be described as 'remnant riparian' vegetation. Findings have revealed that climes with well-formulated regulation for watershed and riparian protection would have classified such remnant vegetation as either "Endangered" or "Of Concern" e.g. Vegetation Management Act 1999 [71]. This brings to bear the need to urgently address the regulatory deficit for riparian protection in developing countries, to save remnant riparian vegetation and foster its restoration by reducing anthropogenic assault.

The occurrence and association of OCPs in sediments with each riparian station indicate that smaller buffer-areas experienced greater sediment mobility, and thus more transport of hydrophobic/low labile pesticides. Riparian vegetation plays an important role in slowing down the flow velocity, decreasing erosion and stabilizing stream banks. In addition to the bank stabilization offered by well-developed root systems, the amount of vegetation present in the riparian region often influence other riparian system functions [40, 72]. Aside from trapping sediment and nutrients, the presence of riparian vegetation protects soil compartments from erosion, compaction, evaporation as well as reduce water and soil temperature [13]. In terms of ecosystem services, riparian rich areas typically provide several important functions, including water purification from riparian vegetation, improved water quality, improved water esthetics for visitors and also improve the abundance of native fish and wildlife; which not only enhance ecological health but also have financial and economic benefits [40, 73]. Howbeit, the urgency for appropriate management of riparian areas will be a product of the information on their current condition and health.

4.5 Socio-economic implication of riparian area loss

The degradation of river basins and watersheds as seen in the Ikpoba river carries tangible economic, fiscal and social costs because the cost of environmental degradation goes hand in hand with the adverse health and lowering the productivity of citizens. The increased incidence of pesticides in surface water could result in acute pesticide poisoning for consumers increasing the likelihood of morbidity and mortality that could debilitate a significant part of the community workforce [74]. However, improving the quality of a deteriorated Ikpoba river system via appropriate policy and sufficient resource allocation will significantly improve the economic and fiscal future of the Ikpoba area and Edo state at large.

5. Conclusion

This study used the agrarian riparian area of Ikpoba river to demonstrate the capacity of riparian vegetation to protect adjacent river systems from pesticide-laden runoff. The spatial distribution of OCPs in water, sediment and fish matrices were consistent with the relative depletion of riparian vegetation and buffer-width across the stations studied. It is a first report to demonstrate within any sub-Saharan area that, buffer-width and density of vegetation are critical aspects of the filtering capacity of riparian areas. While this study highlights dense riparian vegetation and wide buffer-width of 120 m as suitable for protection of surface water quality and aquatic species, more studies of this nature but of larger scope are recommended to inform contextual management of tropical agrarian watersheds. It is also recommended that extant regulatory instruments be reviewed to reflect strategies and practices needed to improve the ecological functions of existing riparian areas and encourage the restoration of depleted riparian areas. Focused action on addressing policy issues and regulatory gaps on watershed protection are necessary to change the current trajectory of deteriorating river basins within urban African areas. Also, strengthening regulatory requirements for pollutant-filtering buffer zones along agricultural catchments to limit risks to adjacent surface water.

Competing interests' statement

The authors declare that they have no conflict of interest.

Funding

The authors declare no funding was received for this research.

A. Appendix I

Overall geospatial methodological scheme of the study.

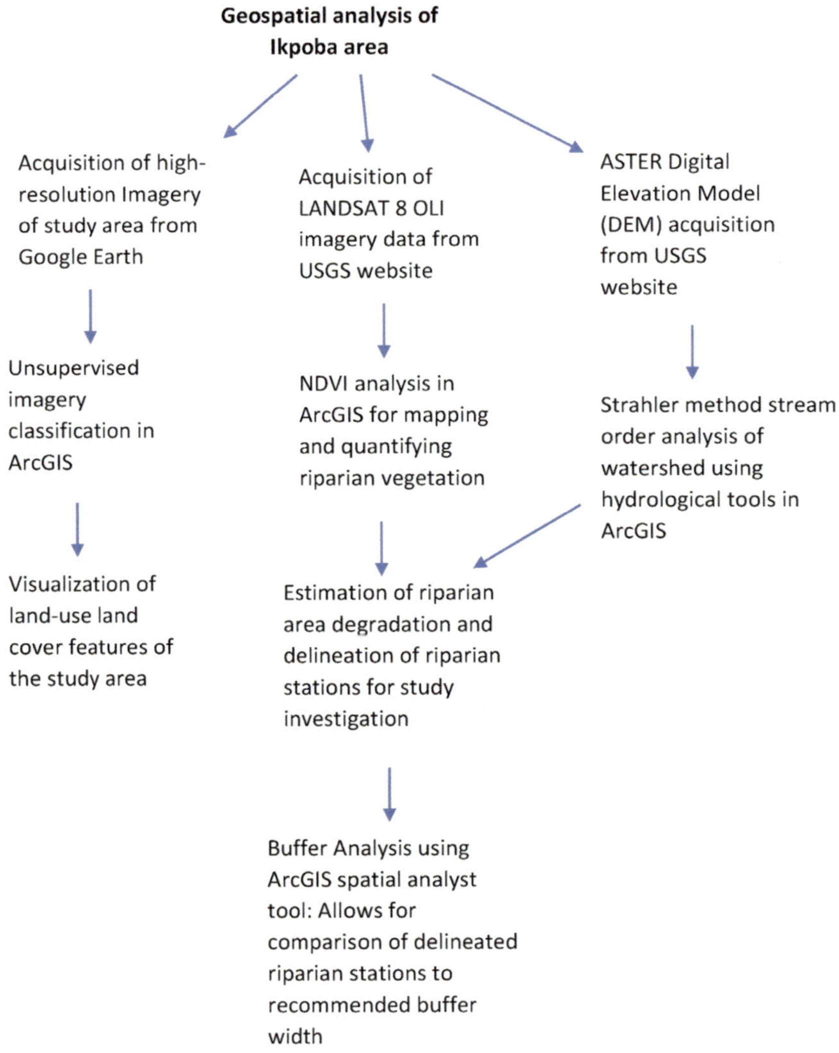

B. Appendix II (RIPARIAN STATIONS)

(d)

(e)

(f)

(a) Full riparian buffer on both sides of river (b) sparse riparian cover on one side of the river with agricultural activity (c) Sparse riparian cover on one side of the river (d) moderate riparian vegetation cover on both sides of the river (e) moderate riparian vegetation cover on both sides of the river (f) sparse riparian cover on both sides of the river

C. Appendix III

Normalized difference vegetation index (NDVI) for predetermined riparian stations within the Ikpoba watershed.

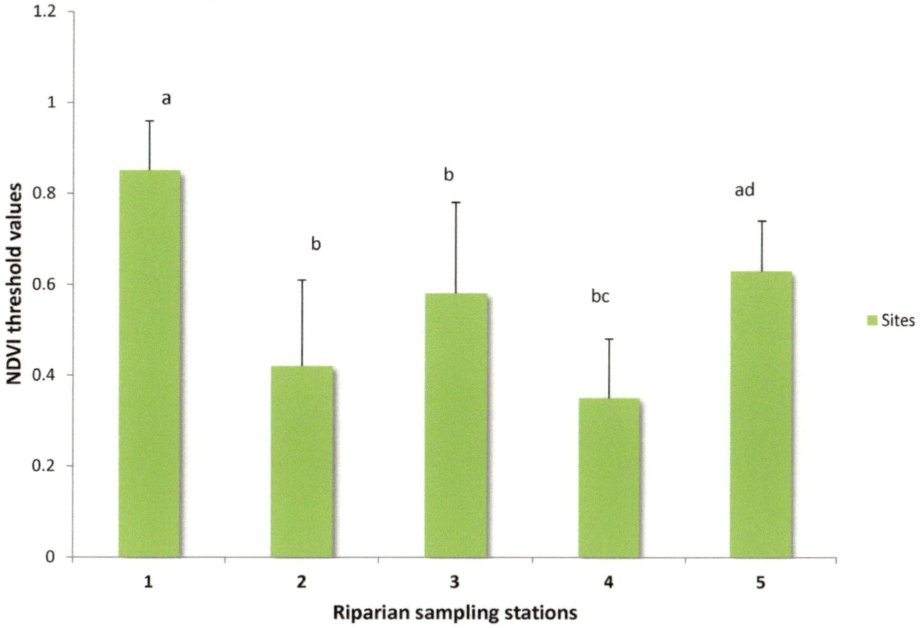

*Stations with the same alphabet are not significantly different

D. Appendix IV

n-octanol–water partition coefficient (log K_{ow}) of pesticides detected within the Ikpoba river catchment.

Pesticide	Log K_{ow}
γBHC	3.73
αBHC	3.8
βBHC	3.81
Endosulfan-I	4.1
Endosulfan-II	4.1
Endos_aldehyde	4.1
Endos_sulfate	4.1
Hept_Epoxide	5
Endrin	5.06

Pesticide	Log K_{ow}
Dieldrin	5.37
Chlordane	5.5
Heptachlor	6.26
Aldrin	6.5
DDT	6.53

Author details

Azubuike Victor Chukwuka[1*] and Ozekeke Ogbeide[2,3]

1 Environmental Quality Control Department, National Environmental Standards and Regulations Enforcement Agency (NESREA), Nigeria

2 Ecotoxicology Research Laboratory, Afromontane Research Unit, Department of Zoology and Entomology, University of the Free State (Qwa qwa Campus), South Africa

3 Environmental Management and Toxicology, University of Benin, Nigeria

*Address all correspondence to: zubbydoo@gmail.com

References

[1] Burdon, F.J., et al., *Assessing the Benefits of Forested Riparian Zones: A Qualitative Index of Riparian Integrity Is Positively Associated with Ecological Status in European Streams.* Water, 2020. **12** (4): p. 1178.

[2] Alemu, T., et al., *Effect of riparian land use on environmental conditions and riparian vegetation in the east African highland streams.* Limnologica, 2017. **66**: p. 1–11.

[3] Hawes, E. and M. Smith, *Riparian buffer zones: Functions and recommended widths.* Eightmile River Wild and Scenic Study Committee, 2005. **15**: p. 2005.

[4] Hansen, B., et al., *Minimum width requirements for riparian zones to protect flowing waters and to conserve biodiversity: a review and recommendations.* Monash University, Melbourne, 2010.

[5] Luke, S.H., et al., *Riparian buffers in tropical agriculture: Scientific support, effectiveness and directions for policy.* Journal of Applied Ecology, 2019. **56**(1): p. 85–92.

[6] Valera, C.A., et al., *The buffer capacity of riparian vegetation to control water quality in anthropogenic catchments from a legally protected area: A critical view over the Brazilian new forest code.* Water, 2019. **11**(3): p. 549.

[7] Poole, K.E. and J.A. Downing, *Relationship of declining mussel biodiversity to stream-reach and watershed characteristics in an agricultural landscape.* Journal of the North American Benthological Society, 2004. **23**(1): p. 114–125.

[8] Owokotomo, A.I., et al., *Watershed land use, surface water vulnerability and public health risks of two urban rivers, Ado-Ekiti, South-West Nigeria.* SN Applied Sciences, 2020. **2**(11): p. 1–21.

[9] Ogbeide, O., I. Tongo, and L. Ezemonye, *Risk assessment of agricultural pesticides in water, sediment, and fish from Owan River, Edo State, Nigeria.* Environmental monitoring and assessment, 2015. **187**(10): p. 654.

[10] Chukwuka, A., O. Ogbeide, and G. Uhunamure, *Gonad pathology and intersex severity in pelagic (Tilapia zilli) and benthic (Neochanna diversus and Clarias gariepinus) species from a pesticide-impacted agrarian catchment, south-south Nigeria.* Chemosphere, 2019. **225**: p. 535–547.

[11] Ogbeide, O., et al., *Relationship between geosorbent properties and field-based partition coefficients for pesticides in surface water and sediments of selected agrarian catchments: Implications for risk assessment.* Journal of Environmental Management, 2018. **217**: p. 23–37.

[12] Lovell, S.T. and W.C. Sullivan, *Environmental benefits of conservation buffers in the United States: evidence, promise, and open questions.* Agriculture, Ecosystems & Environment, 2006. **112** (4): p. 249–260.

[13] Aguiar, T.R., et al., *Riparian buffer zones as pesticide filters of no-till crops.* Environmental Science Pollution Research, 2015. **22**(14): p. 10618–10626.

[14] Grismer, M.E., *Vegetative filter strips for nonpoint source pollution control in agriculture*. 2006: UCANR Publications.

[15] Mohamad Khawlie, et al., *Watershed Characteristics, Landuse and fabric: The Application of Remote sensing and Geographic Information System.* Lakes and Reserviors:Research and Management, 2005. **10**(2): p. 85–92.

[16] Duro, D.C., S.E. Franklin, and M.G. Dubé, *A comparison of pixel-based and object-based image analysis with selected machine learning algorithms for the classification of agricultural landscapes using SPOT-5 HRG imagery.* Remote

sensing of environment, 2012. **118**: p. 259–272.

[17] Qiong, H., et al., *Exploring the use of Google Earth imagery and object-based methods in land use/cover mapping.* Remote Sensing, 2013. **5**(11): p. 6026–6042.

[18] Akasheh, O.Z., C.M. Neale, and H. Jayanthi, *Detailed mapping of riparian vegetation in the middle Rio Grande River using high resolution multi-spectral airborne remote sensing.* Journal of Arid Environments, 2008. **72**(9): p. 1734–1744.

[19] Mendez-Estrella, R., J. Romo-Leon, and A. Castellanos, *Mapping changes in carbon storage and productivity services provided by riparian ecosystems of semi-arid environments in northwestern mexico.* ISPRS International Journal of Geo-Information, 2017. **6**(10): p. 298.

[20] Onyia, N., H. Balzter, and J.-C. Berrio, *Normalized Difference Vegetation Vigour Index: A New Remote Sensing Approach to Biodiversity Monitoring in Oil Polluted Regions.* Remote Sensing, 2018. **10**(6): p. 897.

[21] Ogbodo, S.G., *Environmental protection in Nigeria: Two decades after the Koko incident.* Annual Survey of International Comparative Law, 2009. **15**: p. 1.

[22] Longe, E., et al., *Water resources use, abuse and regulations in Nigeria.* Journal of Sustainable Development in Africa, 2010. **12**(2): p. 35–44.

[23] Hamilton, L., *Red flags of warning in land clearing*, in *Forests, water people in the humid tropics: past, present future hydrological research for integrated land water management. International Hydrology Series. Cambridge University Press, Cambridge*. 2004. p. 866–880.

[24] Zari, M.P., *Ecosystem services analysis in response to biodiversity loss caused by the built environment.* SAPI EN. S. Surveys Perspectives Integrating Environment Society Natural Resources, 2014(7.1).

[25] Amoros, C. and G. Bornette, *Connectivity and biocomplexity in waterbodies of riverine floodplains.* Freshwater biology, 2002. **47**(4): p. 761–776.

[26] Grill, G., et al., *An index-based framework for assessing patterns and trends in river fragmentation and flow regulation by global dams at multiple scales.* Environmental Research Letters, 2015. **10**(1): p. 015001.

[27] Thoms, M.C., M. Southwell, and H. M. McGinness, *Floodplain–river ecosystems: Fragmentation and water resources development.* Geomorphology, 2005. **71**(1–2): p. 126–138.

[28] Ogbeide, O., et al., *First Report on Probabilistic risk assessment of pesticide residues in a riverine ecosystem in South-South Nigeria.* Chemosphere, 2019. **231**: p. 546–561.

[29] Tongo, I., et al., *Levels, distribution and human health risk assessment of organochlorine pesticide residues in surface water from Ikpoba River, Nigeria.* Nigerian Journal of Scientific Research, 2014. **13**(1): p. 26–34.

[30] Victor, R. and A.E. Ogbeibu, *Macrobenthic invertebrates of a stream flowing through farmlands in southern Nigeria.* Environmental Pollution Series A, Ecological and Biological, 1985. **39**(4): p. 337–349.

[31] Tawari-Fufeyin, P. and S.A. Ekaye, *Fish species diversity as indicator of pollution in Ikpoba River, Benin City, Nigeria.* Reviews in Fish Biology and Fisheries, 2007. **17**(1): p. 21–30.

[32] Kamadjeu, R., *Tracking the polio virus down the Congo River: a case study on the use of Google Earth™ in public health planning and mapping.*

International Journal of Health Geographics, 2009. **8**(1): p. 1–12.

[33] Mering, C., J. Baro, and E. Upegui, *Retrieving urban areas on Google Earth images: application to towns of West Africa.* International Journal of Remote Sensing, 2010. **31**(22): p. 5867–5877.

[34] Yu, L. and P. Gong, *Google Earth as a virtual globe tool for Earth science applications at the global scale: progress and perspectives.* International Journal of Remote Sensing, 2012. **33**(12): p. 3966–3986.

[35] Hu, Q., et al., *Exploring the use of Google Earth imagery and object-based methods in land use/cover mapping.* Remote Sensing, 2013. **5**(11): p. 6026–6042.

[36] Aronoff, S., *Remote sensing for GIS Managers ESRI Press*. 2005, Redlands, California: Esri Press 524.

[37] Quinn, P., et al., *The prediction of hillslope flow paths for distributed hydrological modelling using digital terrain models.* Hydrological processes, 1991. **5**(1): p. 59–79.

[38] Sridhar, P., et al., *Prioritisation of miniwatersheds based on Morphometric analysis using GIS.* International Journal of Geomatics Geosciences, 2012. **3**(2): p. 314–320.

[39] Tarboton, D.G., R.L. Bras, and I. Rodriguez-Iturbe, *On the extraction of channel networks from digital elevation data.* Hydrological processes, 1991. **5**(1): p. 81–100.

[40] Pert, P.L., et al., *A catchment-based approach to mapping hydrological ecosystem services using riparian habitat: A case study from the Wet Tropics, Australia.* Ecological Complexity, 2010. **7**(3): p. 378–388.

[41] NRC, N.R.C., *Riparian areas: functions and strategies for management*. 2002: National Academies Press.

[42] Ezemonye, L., T. Ikpesu, and I. Tongo, *Distribution of lindane in water, sediment, and fish from the Warri River of the Niger Delta, Nigeria.* Archives of Industrial Hygiene and Toxicology, 2008. **59**(4): p. 261–270.

[43] Hladik, M. and M.M. McWayne, *Methods of analysis-Determination of pesticides in sediment using gas chromatography/mass spectrometry*. 2012, US Geological Survey.

[44] Steinwandter, H., *Development of microextraction methods in residue analysis.* Emerging strategies for pesticide analysis, 1992: p. 3–50.

[45] USEPA, U.S.E.P.A., *Pesticides in water, soil, sediment, biosolids, and tissue by HRGC/HRMS. EPA Method 1699*, in *EPA-821-R-08-001.* . 2007, U.S. Environmental Protection Agency, Office of Water. : Washington, DC:.

[46] Legendre, P. and L. Legendre, *Numerical Ecology Elsevier*. 1998, Amsterdam, The Netherlands. 853.

[47] Swan, A.R. and M. Sandilands. *Introduction to geological data analysis*. in International Journal of Rock Mechanics and Mining Sciences and Geomechanics Abstracts. 1995.

[48] Garcés-Restrepo, M.F., J.N. Pauli, and M.Z. Peery, *Natal dispersal of tree sloths in a human-dominated landscape: Implications for tropical biodiversity conservation.* Journal of applied ecology, 2018. **55**(5): p. 2253–2262.

[49] Weissteiner, C.J., et al., *An indicator to map diffuse chemical river pollution considering buffer capacity of riparian vegetation—A pan-European case study on pesticides.* Science of the total environment, 2014. **484**: p. 64–73.

[50] Chen, W., et al., *A pesticide surface water mobility index and its relationship with concentrations in agricultural drainage watersheds.* Environmental

Toxicology Chemistry: An International Journal, 2002. **21**(2): p. 298–308.

[51] Heartsill-Scalley, T. and T. Aide, *Riparian vegetation and stream condition in a tropical agriculture–secondary forest mosaic.* Ecological Applications, 2003. **13** (1): p. 225–234.

[52] Shivoga, W.A., et al., *Influences of land use/cover on water quality in the upper and middle reaches of River Njoro, Kenya.* Lakes & Reservoirs: Research & Management,, 2007. **12**(2): p. 97–105.

[53] Anbumozhi, V., J. Radhakrishnan, and E. Yamaji, *Impact of riparian buffer zones on water quality and associated management considerations.* Ecological Engineering, 2005. **24**(5): p. 517–523.

[54] Ogbeide, O., et al., *Comparative gill and liver pathology of tilapia zilli, Clarias gariepinus and Neochanna diversus in owan river (Nigeria): Relative ecological risks of species in a pesticide-impacted river.* Chemosphere, 2019. **234** p. 1–13.

[55] Lorion, C.M. and B.P. Kennedy, *Riparian forest buffers mitigate the effects of deforestation on fish assemblages in tropical headwater streams.* Ecological Applications, 2009. **19**(2): p. 468–479.

[56] Biswas, S.R. and A.U. Mallik, *Disturbance effects on species diversity and functional diversity in riparian and upland plant communities.* Ecology, 2010. **91**(1): p. 28–35.

[57] Pearson, R.G. and N. Stork, *Catchment to reef: water quality and ecosystem health in tropical streams*. 2008, Blackwell Publishing.

[58] Connolly, N.M. and R.G. Pearson, *The effect of fine sedimentation on tropical stream macroinvertebrate assemblages: a comparison using flow-through artificial stream channels and recirculating mesocosms.* Hydrobiologia, 2007. **592**(1): p. 423–438.

[59] Piper, W.H., *Making habitat selection more "familiar": a review.* Behavioral Ecology and Sociobiology, 2011. **65**(7): p. 1329–1351.

[60] Hoover, J.P., *Decision rules for site fidelity in a migratory bird, the prothonotary warbler.* Ecology, 2003. **84** (2): p. 416–430.

[61] Wolf, M., et al., *The attraction of the known: the importance of spatial familiarity in habitat selection in wapiti Cervus elaphus.* Ecography, 2009. **32**(3): p. 401–410.

[62] Schmidt, K.A., *Site fidelity in temporally correlated environments enhances population persistence.* Ecology Letters, 2004. **7**(3): p. 176–184.

[63] Itakura, H., et al., *Site fidelity, diel and seasonal activities of yellow-phase Japanese eels (Anguilla japonica) in a freshwater habitat as inferred from acoustic telemetry.* Ecology of freshwater fish, 2018. **27**(3): p. 737–751.

[64] Norman, L., et al., *Remote sensing analysis of riparian vegetation response to desert marsh restoration in the Mexican Highlands.* Ecological engineering, 2014. **70**: p. 241–254.

[65] Wilson, N.R. and L.M. Norman, *Analysis of vegetation recovery surrounding a restored wetland using the normalized difference infrared index (NDII) and normalized difference vegetation index (NDVI).* International Journal of Remote Sensing, 2018. **39**(10): p. 3243–3274.

[66] McDonough, O.T., et al., *Surface hydrologic connectivity between Delmarva Bay wetlands and nearby streams along a gradient of agricultural alteration.* Wetlands, 2015. **35**(1): p. 41–53.

[67] Tabacchi, E., et al., *Impacts of riparian vegetation on hydrological processes.* Hydrological processes, 2000. **14**(16–17): p. 2959–2976.

[68] Keesstra, S., et al., *Assessing riparian zone impacts on water and sediment movement: a new approach.* Netherlands Journal of Geosciences, 2012. **91**(1–2): p. 245–255.

[69] Fritz, K.M., et al., *Physical and chemical connectivity of streams and riparian wetlands to downstream waters: a synthesis.* JAWRA Journal of the American Water Resources Association, 2018. **54**(2): p. 323–345.

[70] Coleman, D., C. MacRae, and E.D. Stein, *Effect of increases in peak flows and imperviousness on the morphology of southern California streams.* Southern California Coastal Water Research Project Technical Report, 2005. **250**.

[71] Evans, M.C., *Deforestation in Australia: drivers, trends and policy responses.* Pacific Conservation Biology, 2016. **22**(2): p. 130–150.

[72] Easson, G. and L.D. Yarbrough, *The effects of riparian vegetation on bank stability.* Environmental Engineering Geoscience, 2002. **8**(4): p. 247–260.

[73] Loomis, J., et al., *Measuring the total economic value of restoring ecosystem services in an impaired river basin: results from a contingent valuation survey.* Ecological economics, 2000. **33**(1): p. 103–117.

[74] Mateo-Sagasta, J., et al., *Water pollution from agriculture: a global review. Executive summary*. 2017: Rome, Italy: FAO Colombo, Sri Lanka: International Water Management Institute (IWMI). CGIAR Research Program on Water, Land and Ecosystems (WLE).

Section 3

Fluvial Processes, Adaptation and Restoration Strategies

Chapter 7

Morphoanatomical Characteristics in Riparian Vegetation and Its Adaptative Value

Marina García and Damelis Jáuregui

Abstract

Riparian vegetation comprises plant communities that grow laterally to rivers and streams. They have multiple adaptations, which allows them to persist in these variable and dynamic habitats. This chapter focuses on the morphological and anatomical adaptations of vegetative organs, due to the fact that they are more vulnerable to environmental changes that occur in riparian ecosystems. We also discuss some dispersal mechanisms in riparian species exposed to flooding conditions. Most morphoanatomical adaptations in riparian plants reflect constraints imposed by long periods of waterlogging or complete submergence, as well as the high diversity of strategies that species have developed in order to cope with flooding. Furthermore, riparian ecosystems are being impacted by an increasing artificialization of rivers and banks with losses, or profound changes, in the natural riparian vegetation a problem that will increase with the ongoing climate change, and which must be contained. In order to reduce the vulnerability of these ecosystems, a deeper knowledge of the morphoanatomical attributes that make possible the successful adaptation of riparian flora is necessary so as to implement appropriate measures for the rehabilitation and sustainability of riparian ecosystems.

Keywords: riparian vegetation, morphoanatomy, vegetative organs, diaspores, plasticity

1. Introduction

Riparian ecosystems are diverse, dynamic and complex habitats. They are the interface between terrestrial and aquatic systems, encompassing different environmental gradients, communities and ecological processes [1]. Riparian vegetation has ecological functions of immense biological importance within these ecosystems, since it provides habitat and food for a wide variety of terrestrial organisms, as well as maintains its stability, by providing a buffer zone that filters sediments, controls nutrients and stabilizes river banks [2], increasing the biological diversity and productivity of their aquatic communities. In a broad context, the term riparian refers to the biotic communities and the environment adjacent to streams, rivers, ponds, lakes, and some wetlands [3]. Recently, riparian vegetation has been defined as the complex of plant communities growing in the riparian zone that is, in the transitional region between aquatic and terrestrial ecosystems [4].

Riparian vegetation is strongly affected by fluvial processes such as flooding and alluvial soil deposition forming a terrestrial flora that is distinctive in structure and function [2]. The plant diversity of riparian forests comprises a wide range of taxonomic groups and life forms with a variety of morphological, anatomical, physiological and reproductive adaptations which allows them to persist in these habitats [5]. It has been widely accepted that the main threat to the diversity and abundance of native flora of riparian ecosystems is related to anthropogenic activity [6]. This includes the conversion of natural lands into areas of varied anthropic use such as agricultural use, transformation into pastures, introduction of forest species and deforestation practices, among others [1, 7, 8]. This situation, coupled with climate change, has strongly reduced the number of native species in these ecosystems putting many of them at risk of extinction and it is predicted that the impact of climate change on the stressors of riparian communities will probably increase in the future [9].

In order to control the erosion of riparian vegetation, it is necessary to implement plans to promote the natural regeneration and restoration of disturbed areas [10]. This can be possible by knowing the plasticity of the species, that is, their ability to change their phenotype in response to variations in the environment [11]. From the ecological point of view, the plasticity can facilitate the adaptation of species that grow in riparian ecosystems [12], as it allows them to explore different habitats and expand their geographical distribution [13]. However, genetic adaptation is required for the persistence of these characters [14].

This chapter summarizes a wide range of literature regarding the main morphological and anatomical characteristics of vegetative organs, as well as dispersal mechanisms that allow riparian vegetation to adapt and survive in riparian ecosystems. The aim of it is to provide a deep and updated overview of these basic principles, which must be taken into account in order to undertake appropriate measures for the rehabilitation, adaptation and sustainability of riparian ecosystems.

2. Morphological and anatomical adaptations in riparian vegetation

Phenotypic plasticity has long been recognized as the property that allows species to face the heterogeneity of the environment, being of great importance from the adaptive point of view. Due to the fact that it can cause changes in attributes of ecological importance and in heterogeneous environments, species with greater phenotypic plasticity can form ecotypes [11] reducing their extinction risks. Likewise, modular plasticity is also important, since it can influence the activity of meristems and the morphological and anatomical characteristics of vegetative organs mainly stems and leaves more than of the whole plant [15].

Globally, natural and anthropogenic stressors cause multiple disturbances in all riparian environments affecting biological processes in riparian vegetation that reduces their resilience and genetic diversity [6] as summarized in **Figure 1**. One of the factors that most affects the development of riparian plants is being subject to dry soils during some time of the year and flooded soils during others. Another group of plants may be under stress from excessive water during great part of their life cycle because of permanent flooding. In both groups the sheets of water in the soil can cause severe hypoxia, and even anoxic conditions in the radical zone [16].

Riparian plants can cope with the physical disturbances of floods through traits that allow the survival of individuals or that facilitate their rapid growth in the population [17]. The following two main strategies for the adaptation of plants to immersion have been recognized: an escape strategy and a waiting-quiescence one.

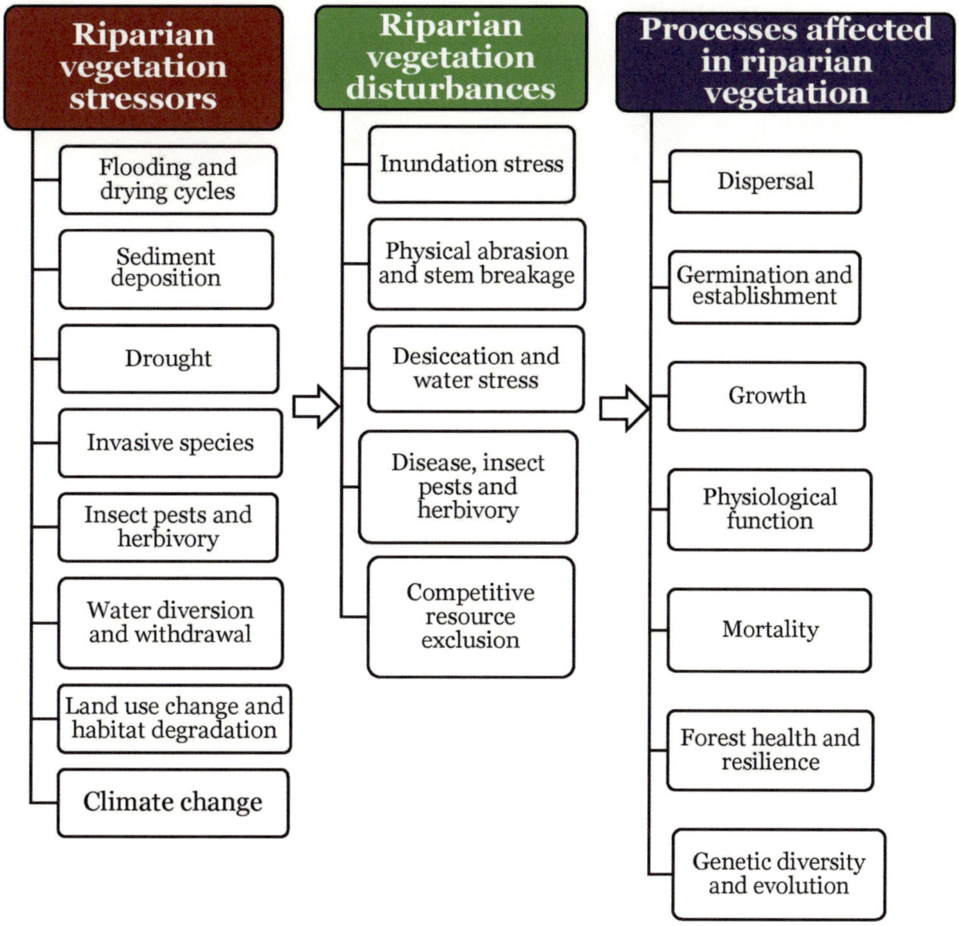

Figure 1.
Main stressors, plant disturbances and processes affected in riparian vegetation.

The former involves the lengthening of shoots to restore the contact of leaves with the atmosphere which is beneficial when a flood is prolonged and superficial. The latter strategy is based on the conservation of energy to prolong survival during the submerging and allowing growth to resume after water backs off, which is beneficial in case of short duration floods and/or deep submersion [18].

The responses of plants to floods are morphoanatomical, physiological and biochemical. These traits vary among species due to their genetic characteristics the age of the plants and the duration and depth of the water sheet to which they are submitted. It has been suggested that morphoanatomical changes are likely involved with long-term acclimation [18].

Resistance strategies include flexibility of stems leaves and roots and the reduction of the size of the plant and brittle branches, among others. These strategies help plants face events that disturb their development [17, 19]. The appearance or modification of morphonatomical characters that enhance oxygen supply could be one of the causes of acclimation of a species under flood conditions [20]. Among the used strategies, the following stand out: the formation of adventitious roots, the development of aerenchyma in submerged roots and stems, the establishment of barriers to prevent the radial loss of oxygen in the roots, the formation of hypertrophic lenticels and cracked stems and the hypertrophy in the stem base [16, 17]. The morphological and anatomical adaptations in the root and shoot of riparian vegetation are summarized in **Figure 2**.

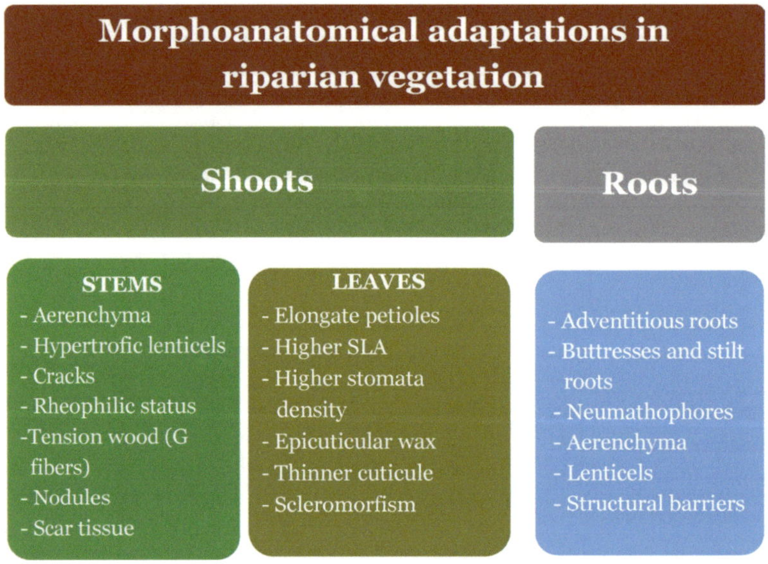

Figure 2.
Main morphoanatomical adaptations in riparian vegetation.

2.1 Morphoanatomical adaptations in the radical system

Flooded soils are low in oxygen, due to reduced exchange rates with the atmosphere. Hypoxia in the radical tissues induces anaerobic respiration and increased consumption of stored carbohydrates [19]. Lack of energy affects growth, cell maintenance, reduced nutrient absorption and even root death [21].

The primary root system of plants subject to flooding is strongly affected by reduction of their biomass, as well as by that of the length and diameter of the roots. This is due to cellular damage at the membrane level. In some species, roots can turn black and die when flooded. In mature trees the root system can break down during prolonged soil flooding, so the production of new roots is an important process [22].

The riparian plants can avoid (or "resist") the effects of water deficit by developing either deep roots that extend to the water table, or resistant traits that allow rapid recolonization [17].

Among the adaptation strategies in the radical system to hypoxic conditions we can mention:

The formation of adventitious roots, which is an adaptive mechanism of ecological importance in riparian plants, since it allows to replace roots that have died or have been affected due to waterlogging. It is necessary to distinguish among different types of adventitious roots, as it has been proven that their formation is regulated differently. The adventitious roots are separated into two large groups: the first one includes those potentially established in the embryo such as those of the monocotyledons, the nodal roots of some eudicotyledons and the stilt roots; the second group is composed of the roots whose formation is induced by a stressor (in this case flooding) which may or may not be nodal [23].

In some species the primordia of adventitious roots are formed during normal development, but they emerge when plants are under a sheet of water (e.g. *Rumex palustris*). However, it has been suggested that in others taxa submergence induces root development (e.g. *Eucalyptus* species). The moment of emergence of adventitious roots induced by floods is specific to each species [22, 24] and depends on factors such as the stage of development of the plant, water temperature, and the depth and duration of the flood [25].

Floods can promote the formation of roots from the hypocotyl, knots beneath the ground (crown), or above it (brace), as well as internodes. Adventitious roots facilitate gas flow, water transport, and nutrient absorption during and after floods ensuring plant survival [26]. Likewise, it has been pointed out that the formation of these roots in trees improves their internal aeration and promotes water absorption after the first floods as observed in a Mapire igapó [20].

The formation or increase in number of adventitious roots has been confirmed by various authors in different tropical riparian species, such as: *Polygonum ferrugineum* and *P. stelligerum* [27], *Senna reticulata* [28], *Tabebuia rosea* [29] and *Triplaris gardneriana* [30].

Another important adaptive trait of roots in riparian vegetation is the formation of abundant buttress roots and stilts [31]. Buttress roots are common in plants that grow on riverbanks and streams, as well as in trees that lack a deep root. They are closely related to the duration of the flood period and the dynamics of the habitat [32]. It has also been pointed out that buttresses roots are common in trees that develop on substrates where their anchorage is difficult, such as those with a thin layer of sediments (e.g. *Byrsonima amazonica*), while in areas with deep layers many stilt roots are formed, as occurs in *Alchornea castaneifolia* [31]. Buttress and arch roots are piles in sloping areas, providing stability. Its occurrence has been cited in *Aquilaria malaccensis* and *Drypetes* spp. [33].

In some riparian species pneumatophores are noticeable, which are specialized roots with negative geotropism that grow outside the water. These develop as ascending erect organs with lenticels along the surface and spongy tissue that allow the flow of oxygen and facilitates its diffusion throughout the plant. This type of roots has been observed in *Pithecellobium latifolium* and, experimentally, in some palms that grow at the headwaters of rivers and in swampy areas, for example, *Euterpe* and *Mauritia* [31].

Another root adaptation to flooding is the increase in its porosity generated by evenly distributed intercellular spaces, small lagoons or the formation of aerenchyma in the cortex [23, 34]. This tissue constitutes a low-resistance internal pathway for the movement of gases among the different parts of the plants thus improving oxygen supply to the roots [35]. The presence of this tissue prevents anaerobiosis in the root system, making it an efficient mechanism that contributes to the general adaptation of tree species to long-term floods [36].

It has been indicated that roots of plants in mesic environments have a porosity (percentage of air spaces) of 2 to 7% of their volume, while in flooded areas this porosity can reach values close to 60% [37]. The formation of aerenchyma may be less important for the longitudinal flow of oxygen in adult trees because lagoons are destroyed [38]. In this case, ATP production in radical cells is achieved by reducing the number of cells that consume oxygen in the cortex [39].

Species that normally grow in the Amazon were classified according to the presence and development of gas exchange and mobilization system: the first group with roots lacking spaces; the second group with pronounced intercellular spaces in roots, but which are not modified with hypoxia treatments; the third group with intercellular spaces in roots, which are partially modified when plants are subject to flooding; and the last group in which species produce a large quantity of adventitious roots with well-developed aerenchyma [34].

In herbaceous species or in seedlings of different biotypes primary aerenchyma of diverse origin occurs. However, when secondary growth has occurred secondary aerenchyma is formed, mainly from the phellogen [35, 40].

The presence of constitutive aerenchyma, or that induced by hypoxia in roots and stems of riparian plants, has been verified in various studies. The first occur in *Guazuma ulmifolia* [41] and in three shrub species of Melastomataceae that inhabit areas with frequent flooding [42], and the last in *Eucalyptus camaldulensis subsp. refulgens* an Australian timber species which grows mainly on the river banks and their alluvial plains, but which can endure drought [22].

Likewise, in *Rumex palustris*, a species tolerant to flooding, has been indicated a greater development of the aerenchyma than in *R. acetosa*, the latter being an intolerant species to this condition [43]. In *Tabebuia rosea* seedlings, aeriferous parenchyma was evident in roots and stems, as well as in *Handroanthus chrysotrichus* stems in which aerenchymatous phellem was noticed [29].

The development of aeriferous parenchyma, lenticels and fissures or cracks allows the axial diffusion of gases especially oxygen between air space and the internal part of roots [34].

Deposition of hydrophobic compounds (suberin and lignin) can be found in the cortical region of the roots as barriers of thick-walled cells. Suberin deposits are able to prevent radial loss of oxygen from the root cortex to the rhizosphere [17, 34, 44]. Suberized walls are reported to diverse degrees, mainly in the exodermis of young roots of species typical of varzeas in Manaus, Brazil [39].

The hypoxic condition induced by floods promotes a greater diameter in roots due to a thicker cortex. Likewise, the area occupied by the stela is smaller which suggests that both features are variables to consider in the adaptability to waterlogging conditions [45]. However, this response may be the opposite, and a reduction in root diameter may occur [29, 46]. In *Guazuma ulmifolia* and *Genipa americana*, species with high plasticity adaptable to flooded soils, the reduction on these traits is associated with energy savings [41].

Previous considerations correspond mainly to the adaptations observed in root system of herbaceous plants and seedlings of tree species, which could be associated to the difficulty of working with underground systems in trees. It is important to further study the roots traits of adult individuals since it is not certain that those obtained from seedlings could be extrapolated [20, 34].

2.2 Morphoanatomical adaptations in the shoots

Initially, the importance of some biotypes in riparian vegetation should be highlighted. Trees and shrubs play an important role by blocking wind and stabilizing terrain. Herbs contribute to the stabilization of soil and are valuable tools for the rehabilitation of degraded riparian environments [47]. It is important to know the mechanisms that each species has to tolerate or adapt to the conditions of each particular habitat in order to have tools to choose useful species to reforest when necessary.

The impact of hypoxia on stem tissues has not been widely studied. It was also pointed out that it occurs particularly at the meristem level, especially when flood water is muddy and makes it difficult for light to pass through [21].

2.2.1 Stems

Few details have been reported regarding the adaptive importance of rhizomes and stolons in plants from riparian environments. However, it is known that amphibian species can develop rhizomatous and stoloniferous stems with cortex made up of aerenchyma, which constitutes an adaptive trait. These types of stem have been observed in amphibian species such as *Cynodon dactylon* and *Paspalum distinum* [48].

Likewise, the shrubby species *Ficus squamosa* is able to grow stolon-like stems when it grows on banks of the Ping River in Thailand [49]. The development of these types of stems constitutes an adaptive advantage not only because the presence of aerenchyma allows them to stay afloat, but also because they constitute diaspores of propagation of the species since these fragments can be part of a new individual [48].

Stem nodulation has been observed in several species of legumes that inhabit flooded, or likely to flood places. This phenomenon is an adaptation which allows legumes to fix nitrogen in these environments. Plant species that exhibit stem nodulation are typically tropical or subtropical and grow in wetlands, rivers or lake margins, and belong to *Aeschynomene*, *Sesbania*, *Neptunia* and *Discolobium* genera [50]. Along with the formation of nodules, some species develop a large number of parenchymal cells, which facilitates the entry of sufficient oxygen for different metabolic functions [51].

Stems do not have selective barriers (such as the exodermis and endodermis) but they can develop a cuticle, which due to its hydrophobic characteristics can perform the same function as the previously mentioned tissues [40]. However, it is possible that in riparian plants with stolons and rhizomes those tissues differ. Nonetheless, this assumption must be verified through further morphoanatomical studies.

Species adapted to prolonged flooding avoid anoxia by spreading out of the water [17]. The lengthening of seedlings shoots or epicormic is a response to flooding of various plant species. It ensures the restoration of contact with the atmosphere in order to maintain internal aeration [52]. This lengthening occurs mainly in internodes and petioles, which causes leaves to approach the surface achieving better lighting conditions. *Rumex palustris* [53] and *Chloris gayana* benefit from this strategy, which is considered an escape during prolonged periods of flooding [54].

Sometimes fissures or cracks are visible in the basal part of young stems of grasses and trees, which are the result of pressure exerted by the development of cells of the aerenchymatous phellem on the epidermis and on some other external cortical layer until it is broken exposing the internal tissue to the atmosphere [55]. In *Sesbania javanica* [56], *Tabebuia rosea* and *Myracrodruon urundeuva* [29], cracks were observed on the surface of the seedling stems, growing in flooded soils.

Many flood-tolerant riparian species develop hypertrophic lenticels on stems which penetrate the phellogen layer and allow gas exchange. Hypertrophic lenticels facilitate the absorption of oxygen by the plant, but they are also supposed to contribute to the release of carbon dioxide and volatile compounds, such as acetaldehyde and ethanol [57]. These structures were observed in the riparian species *Schinus terebinthifolius*, in which they appear as soft whitish masses in the basal portion of the stems [58]. They are also a response to flood in *Sesbania javanica* [52], *Tabebuia avellanedae*, a riparian species of the Paraná River-Brazil [59], and in *Guazuma ulmifolia* [41]. In *Genipa americana* they are still present under non-stressful conditions, thus they are considered constitutive. However, waterlogging stress causes an increase in its frequency and size [41].

Some species of riparian zones have characteristics in their woods that allow them to adapt to watercourses. Therefore, they constitute an alternative in the stabilization and protection of river slopes. Knowledge of the internal structure of wood is essential to understand its behavior in these habitats, particularly when it comes to structural features related to flexibility [60].

The set of adaptive characteristics of wood can, as a whole, be called "anatomical rheophilic syndrome". It is characterized by: small and solitary vessels of 100 μm in diameter, multiple radial, with walls 2.5–5.0 μm thick; scarce or absent axial parenchyma; narrow rays; and the presence of gelatinous fibers [61]. These woods are constituted by: 10 to 30 percent of vessels, 0 to 5 percent of axial parenchyma, 10 to 20 percent of rays and 60 to 70 percent of fibers. Several of these characteristics have been observed in riparian species such as *Colliguaya brasiliensis* [60] and *Aspidosperma riedelii* [62].

In riparian environments trees can be inclined or crooked, which is the result of the so-called tension wood, characterized anatomically by the lack or scarce lignification in cell walls of the fibers, as well as by the presence of an internal gelatinous layer in them (so-called G fibers). These cells support high stresses in mature organs, allowing their movement, or reinforcing their structure and stability [63]. G fibers have been observed in *Sebastiana commersoniana* [64] and *Ludwigia* species [65] when these grow in flooded soils.

Likewise, the presence of macules in the wood has been highlighted. This structures are parenchymal cells with thicker walls than those of the adjacent cells with starch and phenolic compounds, which facilitate anaerobic survival. These structures were observed in *Sebastiana commersoniana* [64] and in *Eugenia inundata* [66].

2.2.2 Leaves

Leaves are the most plastic plant organ. Therefore, their morphology and anatomy are usually closely linked to the environment in which plants grow [67]. Several foliar parameters have been considered important for the adaptation to specific riparian microenvironments among them size and mass, water content, and anatomical characteristics of the leaf surface and mesophyll.

A common foliar adaptation in riparian species subject to waterlogging conditions is the elongation of the petioles of submerged leaves which allows them to emerge, thus avoiding the low luminosity to which they are exposed during flood periods [52]. In the Amazonian floodplain, many tree species shed their leaves in the first week to months of waterlogging or submergence and produce new ones near the end of the flood period [68]. However, in these ecosystems there are also evergreen trees that keep their leaves throughout the waterlogging period, even when completely submerged. These leaves are generally xeromorphic [20, 69, 70].

In trees of a seasonally flooded forest of the Mapire River in Venezuela, it has been found that both submerged and non-submerged leaves have xeromorphic characteristics such as: thick cuticle, large epidermal cells, bifacial mesophyll and abundance of sclerenchyma. These traits remaining practically unchanged regardless of the phase of the flood cycle [70]. It is believed that these trees experience a water deficit during the waterlogging phase similar to that faced by plants from dry habitats, due to a decrease in water absorption by roots. Therefore, these xeromorphic leaves allow trees to cope with the drought during the non-waterlogging phase [69]. In riparian species that are not periodically flooded both mesomorphic and xeromorphic adaptations may occur [71].

The riparian species *Guazuma ulmifolia* and *Sapium glandulosum* have contrasting histological characteristics in the leaf blade [71] and also in the midrib and petiole (**Figure 3**). *G. ulmifolia* shows higher density of xylem vessels (narrow diameter) and a greater development of sclerenchyma (**Figure 3A, B**) than *S. glandulosum* (**Figure 3C, D**) on both midrib and petiole. These traits are associated with xeromorphic leaves [72, 73] because they can maximize the efficiency of water transport into the leaf and prevent foliar collapse under drought conditions [74]. These anatomical differences can help explain the environments in which they grow: *S. glandulosum* frequently inhabits floodplains and swampy areas [75, 76] whilst *G. ulmifolia* has been reported in both floodplains [77, 78] and dry and intervened areas [71, 79]. Furthermore, the secretory cavities in the midrib and petiole of *G. ulmifolia* (**Figure 3A, B**) accumulate a mucilage that can retain water [41] and has also been associated with a defense mechanism against biotic stressors [80].

The specific leaf area (SLA, $cm^2\ g^{-1}$) is related to the photosynthetic capacity, nitrogen content, longevity of the leaf and foliar architecture [81]. It is usually lower in species growing in arid environments [82] and higher in wetland habitats [83],

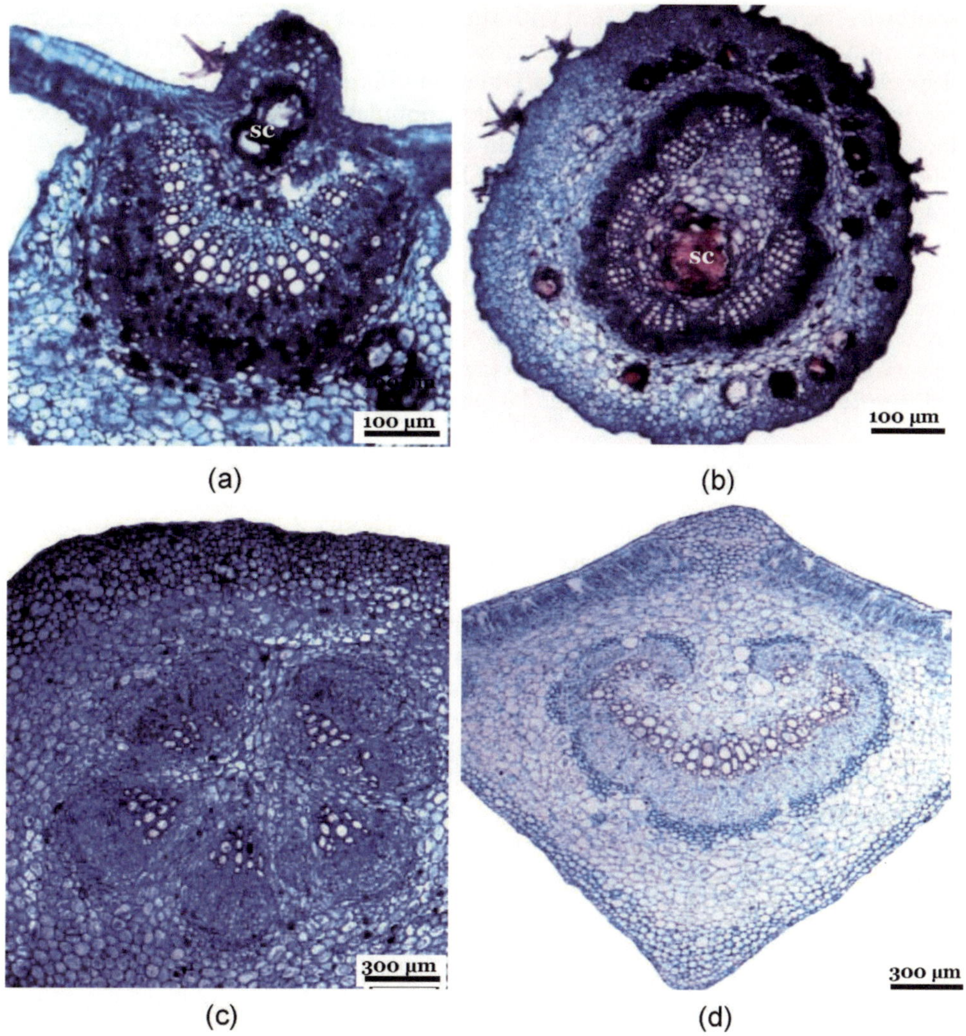

Figure 3.
View of the midrib and petiole cross section in Guazuma ulmifolia *and* Sapium glandulosum. *A-B: G. ulmifolia, midrib (A) and petiole (B); C-D: S. glandulosum, midrib (C) and petiole (D). sc: secretory cavity.*

as it can promote gas exchange in plants whose leaves must withstand waterlogging conditions. However, in trees from a igapó of the Venezuelan Amazon, a decrease in SLA with flooding has been observed in both emerged and submerged leaves [70]. This seems to be associated with an acclimation process that can promote an increase in CO_2 conductance in the foliar mesophyll under waterlogging conditions [20].

In riparian ecosystems from Europe, the black poplar (*Populus nigra*) is one of the most studied species due to disturbances caused by the degradation of its natural habitat. A morphometric analysis on leaves across 17 native populations of *P. nigra* from the banks of six rivers in Croatia and Bosnia and Herzegovina including the hairy type (*P. nigra* subsp. *caudina*) of sub-Mediterranean climate, and the typical poplar type of continental riparian forests showed that that the best features to differentiate these two types of populations were the distance between the broadest portion and the base of the blade together with the length of the petiole [84].

In genotypes of poplar species *P. deltoides*, *P. balsamifera*, *P. angustifolia* and hybrids between them it has been found that all of them have a high density of small stomata [85]. This can be associated with a high stomatal conductance and a lower water use efficiency [86] consistent with their behavior as ecological pioneers that grow rapidly in low fertility areas after floods or other disturbances [87]. Likewise, all species had abundant cuticular wax which would reduce transpiration in poplars growing in dry riparian regions [88]. In native populations of *P. nigra* of the Serbian alluvial zone, no differences were found in stomatal characteristics which could indicate that each of them has a similar combination of genotypes [89].

Invasive exotic plants have a negative impact on native riparian species, mainly because they affect the flow of energy and the cycling of matter, as well as by their excessive use of resources. A study on *Echinocystis lobata* and *Parthenocissus quinquefolia* (invasive climbing species) growing on *Salix caprea* and *S. fragilis*, respectively, showed that in both willow species the leaf cuticle became thinner relative to that of plants that grew without these vines [90]. This is a disadvantage since the cuticle contributes to the mitigation of the adverse impact caused by both biotic and abiotic stressors.

2.3 Dispersion units

The dispersal of plants in riparian environments is affected by the hydrogeomorphic characteristics of each place. Riparian species can be dispersed by various vectors, including wind (anemochoria), animals (zoocoria), and water (hydrochoria). However, the proportion in which these mechanisms occur is not quantified [17]. Hydrochoria is highlighted as a dominant dispersal vector in riparian habitats [17, 91, 92]. Both vegetative and generative propagules can be dispersed by water [92]. In these environments, the formation of a seed bank is not feasible [19].

Ichthyocoria is very common in tropical riparian environments. Diaspores dispersed by fish are highly variable in shape, texture, color and taste [92]. The intact seeds excreted by fish are small with high nutrient content in the pericarp of the fruits, as occurs in *Pouteria glomerata* [92]. Frugivorous fish can be effective dispersers of large, non-floating seeds, thus contributing to dispersal over long distances as well as upstream [93].

In several species hydrochoria is a secondary dispersal mechanism, as propagules often fall to the ground before being carried away by the water current. Other species drop them directly into the stream (*Disa uniflora*). Diaspores that lack buoyancy can be transported by water by adhering to floating objects such as logs, branches and litter [92].

In riparian plants hydrochoria offers several advantages: 1. Floating hydrochorous can disperse long distances compared to other dispersal vectors. 2. The dispersal distance of floating hydrochorous is largely independent of their mass consequently, they tend to be heavier than those dispersed by wind and heavier seeds produce taller seedlings (with the advantages that this implies). 3. Immersion in water prevents desiccation, thereby increasing the longevity of vegetative propagules. 4. It increases the odds that propagules will be deposited in suitable sites for germination and growth [17].

Small seeds can occupy a greater number of available microsites, while large ones allow a higher rate of seedling development since they have more endosperm or cotyledons with more accumulated substances that favor embryo development [91].

Floating propagules have cork or some spongy tissue and epidermis with waxes or thick cuticle which prevents the imbibition and collapse of them. The seeds of riparian species such as *Annona montana* and *Hevea spruceana* have aeriferous tissue in variable proportions [91]. *Manicaria saccifera* and *Crudia acuminata* have a hollow endosperm, while *Carapa guianensis* and *Pachira aquatica* have sparse endosperms which allow them to float [94]. *E. tenuifolia* has corky tissue and develops spongy tissue in the seed coat. *Tabebuia* species have their coat seed suberous and hydrophobic [95]. Many of these diaspores go through a long period of floating without losing their viability [91].

The presence of spikes, hooks, and other appendages on the surface of diaspores can reduce water surface tension. These structures allow them to attach to leaves, branches, and other surfaces [92]. They also catch air bubbles and, consequently, increase buoyancy [95]. The diaspores can have a hard cover, which gives them greater tolerance to anoxic conditions, resistance to abrasion and tolerance to burial allowing the seed to be transported with sediment and garbage, thus increasing its longevity in the soil [17].

In the Amazon, after fruits ripen which generally occurs when there are high levels of water [95] seeds fall into the water and can float and/or submerge for several weeks without losing their viability (quiescence). Seed germination begins only when the flood recedes, although in some species the radicle may protrude or even the entire seedling may be produced, while floating, as occurs in *Carapa guianensis* [96].

Vegetative fragments can detach and disperse during disturbance events, improving recolonization capacity. These fragments usually float and help the dispersion. Furthermore, they can remain viable for weeks after separation. They can begin to produce roots shortly after detachment, which facilitates their establishment by being immobilized somewhere [97]. *Salix martiana* shows intense vegetative propagation through parts of broken stems. In the case of *Eugenia inundata*, colonizing species of igapó areas in Brazil, lower branches take root [98] so vegetative fragments contribute to the fast and effective establishment in periods when soil lacks water.

3. Conclusions

Riparian areas have been substantially degraded mainly due to anthropogenic factors which have caused erosion of the riparian vegetation. This situation will become more serious in coming years due to events associated with climate change. In this chapter, we have discussed the main morphoanatomical adaptations of the vegetative organs and dispersal propagules of riparian species which determine their ability to survive in such variable and dynamic habitats. In species that face prolonged flooding periods, these adaptations help overcome the anoxia imposed by waterlogging. However, in arboreal species typical features of xeromorphism may be present especially in leaves, which not only allows them to face the water deficit caused by the decrease in absorption of water by roots due to waterlogging, but also drought conditions during the non-waterlogging period. In riparian species from drier areas, both mesomorphic and xeromorphic adaptations may occur. From an ecological point of view, more plastic morphoanatomical adaptations can reduce the risk of extinction of riparian species since this allows them to explore different habitats and thus, expand their geographic distribution. Understanding the scientific principles that support the adaptation of species to riparian habitats is crucial for the decision making process by landscapers and government entities that are responsible to promote the protection and conservation of these ecosystems. In order for this to be achieved, the contribution of different disciplines is required, since interdisciplinary teams are the ones that have the greatest possibility to successfully handle this challenge.

Author details

Marina García[1*] and Damelis Jáuregui[2]

1 Faculty of Agronomic Engineering, Technical University of Manabí (UTM), Portoviejo, Manabí, Ecuador

2 Institute of Agricultural Botany, Agronomy Faculty, Central University of Venezuela, Maracay, Aragua, Venezuela

*Address all correspondence to: marina.garcia@utm.edu.ec

References

[1] Vázquez G, Garcia JG, Franco G, Castillo F, Escobar A, Guillén ML, et al. Ecosistemas ribereños: un paisaje fragmentado. CONABIO. Biodiversitas. 2015;**119**:7-11

[2] Richardson D, Holmes P, Esler K, Galatowitsch S, Stromberg J, Kirkman S, et al. Riparian vegetation: degradation, alien plant invasions, and restoration prospects. Diversity and Distributions. 2007;**13**:126-139. DOI: 10.1111/j.1472-4642.2006.00314.x

[3] Naiman R, Fetherston K, Mckay S, Chen J. Riparian forest. In: Naiman RJ, Bilby RE, editors. River Ecology and Management: Lessons from the Pacific Coastal Ecoregion. New York: Springer-Verlag; 1998. pp. 289-323. DOI: 10.1007/978-1-4612-1652-0_12

[4] Dufour S, Rodríguez-González PM. Riparian zone/riparian vegetation definition: principles and recommendations. Report, Cost Action CA16208 Converges, 2019. p. 1-20. Available from: https://converges.eu/resources/riparian-zone-riparian-vegetation-definition-principles-and-recommendations/

[5] Sarr DA, Odion DC, Hibbs DE, Weikel J, Gresswell RE, Bury BB, Czarnomski NM, Pabst RJ, Shatford J, Moldenke AR. Riparian zone forest management and the protection of biodiversity: A problem analysis. Technical Bulletin N° 908. Research Triangle Park. NCASI; 2005. p. 1-133.

[6] Stella JC, Bendix J. Multiple stressors in riparian ecosystems. In: Sabater S, Elosegi A, Ludwig R, editors. Multiple Stressors in River Ecosystems: Status, Impacts and Prospects for the Future. Amsterdam: Elsevier; 2019. pp. 81-110. DOI: org/10.1016/B978-0-12-811713-2.00005-4

[7] Rodríguez-Olarte D, Barrios Gómez M, Marrero C, Marco L. Río Turbio: un síndrome urbano en la vertiente andina del Orinoco. In: Rodríguez-Olarte D, editor. Ríos en riesgo de Venezuela. Universidad Centro Occidental Lisandro Alvarado: Venezuela. 2017. pp. 59-74

[8] Gamarra-Torres O, Barrena-Gurbillón M, Ordinola-Ramírez C, Barboza-Castillo E, Leiva-Tafur D, Rascón-Barrios J, et al. Calidad del bosque de ribera en la cuenca del río Utcubamba, Amazonas, Perú. Arnaldoa. 2018;**25**:653-678. DOI: org/10.22497/arnaldoa.252.25218

[9] Kiesel J, Guse B, Bormann H. Projecting the consequences of climate change on river ecosystems. In: Sabater S, Elosegi A, Ludwig R, editors. Multiple Stressors in River Ecosystems - Status, Impacts and Prospects for the Future. Amsterdam: Elsevier; 2019. pp. 281-301. DOI: org/10.1016/B978-0-12-811713-2.00016-9

[10] Price P, Tubman W. Structure and characteristics of riparian lands. In: Lovett S, Price P, editors. Principles for Riparian Lands Management, Land & Water: Australia. 2007. pp. 1-12 Available from: www.lwa.gov.au

[11] Gratani L. Plant phenotypic plasticity in response to environmental factors. Advances in Botany. **2014**;**2014**:1-17. DOI: org/10.1155/2014/208747

[12] Naiman RJ, Décamps EH. The ecology of interfaces: Riparian zones. Annual Review of Ecology and Systematics. 1997;**28**:621-658. DOI: 10.1146/annurev.ecolsys.28.1.621

[13] Sultan SE. Promising directions in plant phenotypic plasticity. Perspectives in plant ecology. Evolution and Systematics. 2004;**6**:227-233. DOI: org/10.1078/1433-8319-00082

[14] Richter S, Kipfer T, Wohlgemuth T, Guerrero CC, Ghazoul J, Moser B.

Phenotypic plasticity facilitates resistance to climate change in a highly variable environment. Oecologia. 2012;**169**:269-279. DOI: 10.10007/s0044s-011-2191-x

[15] Kroon H, Huber H, Stuefer JF, Groenendael JM. A modular concept of phenotypic plasticity in plants. New Phytologist. 2005;**166**:73-82. DOI: 10.1111/j.1469-8137.2004.01310.x

[16] Medri ME, Bianchini E, Pimenta JA, Colli S, Müller C. Estudos sobre tolerância ao alagamento em espécies arbóreas nativas da bacia do rio Tibagi. In: Medri ME, Bianchini E, Shibatta OA, Pimenta JA, editors. A bacia do rio Tibagi. Paraná:Londrina; 2002. p. 133-172-2002.

[17] Catford JA, Jansson R. Drowned, buried and carried away: Effects of plant traits on the distribution of native and alien species in riparian ecosystems. New Phytologist. 2014;**204**:19-36. DOI: 10.1111/nph.1295

[18] Voesenek LA, Bailey-Serres J. Flooding tolerance: O2 sensing and survival strategies. Current Opinion in Plant Biology. 2013;**16**:647-653. DOI: 10.1016/j.pbi.2013.06.008

[19] Bendix J, Stella JC. Riparian vegetation and the fluvial environment: A biogeographic perspective. In: Shrode JF, Butler DR, Hupp CR, editors. Treatise on Geomorphology. Vol. 12. Ecogeomorphology, San Diego: Academic Press; 2013. pp. 53-74. DOI: 10.1016/B978-012-374739-6.00322-5

[20] Herrera A. Responses to flooding of plant water relations and leaf gas exchange in tropical tolerant trees of a black-water wetland. Frontiers in Plant Science. 2013;**4**:1-12. DOI: 10.3389/fpls.2013.00106

[21] Pedersen OAD, Perata PB, Laurentius, Voesenek L. Flooding and low oxygen responses in plants. Functional Plant Biology. 2017;**44**:iii–vi. DOI: org/10.1071/FPv44n9_FOses

[22] Argus RE, Colmer TD, Grierson PF. Early physiological flood tolerance is followed by slow post-flooding root recovery in the dryland riparian tree *Eucalyptus camaldulensis* subsp. *refulgens*. Plant Cell Environment. 2015;**38**:1189-1199. DOI: org/10.1111/pce.12473

[23] Steffens B, Rasmussen A. The physiology of adventitious roots. Plant Physiology. 2016;**170**:603-617. DOI: 10.1104/pp.15.01360

[24] Dawood T, Rieu I, Wolters-Arts M, Derksen EB, Mariani C, Visser EJW. Rapid flooding-induced adventitious root development from preformed primordia in *Solanum dulcamara*. AoB Plants 2014;6: plt058. DOI: org/10.1093/aobpla/plt058

[25] Zhang Q, Visser EJW, de Kroon H, Huber H. Life cycle stage and water depth affect flooding-induced adventitious root formation in the terrestrial species *Solanum dulcamara*. Annals of Botany. 2015;**116**:279-290. DOI: org/10.1093%2Faob%2Fmcv095

[26] Sauter M. Root responses to flooding. Current Opinion in Plant Biology. 2013;**16**:282-286. DOI: org/10.1016/j.pbi.2013.03.013

[27] Marques AM, Moscheta IS. Anatomy of the root of eight species of emergent aquatic macrophytes from the upper Paraná river, Paraná State, Brazil floodplain. Acta Scientiarum. Biological Sciences. 2010;**32**:297-304. DOI: 10.4025/actascibiolsci.v32i3.5509

[28] Arenque BC, Grandis A, Pocius O, de Souza AP, Buckeridge MS. Responses of *Senna reticulata*, a legume tree from the Amazonian floodplains, to elevated atmospheric CO_2 concentration and waterlogging. Trees. 2014;**28**:1021-1034. DOI: 10.1007/s00468-014-1015-0

[29] De Oliveira AS, Ferreira CS, Graciano-Ribeiro D, Franco AC. Anatomical and morphological modifications in response to flooding by six Cerrado tree species. Acta Botanica Brasilica. 2015;**29**:478-488. DOI: org/10.1590/0102-33062014abb0035

[30] Pontara V, Bueno ML, Scremin-Dias E. Flooding avoidance *Triplaris gardneriana* Wedd. (Polygonaceae): Growth and morpho-anatomical aspects. Acta Scientiarum. Biological Sciences. 2016;**38**:341-346. DOI: 10.4025/actascibiolsci.v38i3.31267.

[31] Parolin P. Diversity of adaptations to flooding in trees of amazonian floodplains. Pesquisas, Botânica. 2012;**63**:7-28

[32] WittmannF PP. Aboveground roots in Amazonian floodplain trees. Biotropica. 2005;**37**:609-619. DOI: 10.1111/j.1744-7429.2005.00078.x

[33] Hidayat MJ, Abd AK, Maslinda CO, Maimunah DS. Variations of riparian vegetation along the river corridors of Sg Johor. Journal of Advanced Research Design. 2017;**28**:1-11 Available from: www.akademiabaru.com/ard.html

[34] Haase K, Rätsch G. The morphology and anatomy of tree roots and their aeration strategies. In: Junk WJ, Piedade MTF, Wittmann F, Schöngart J, Parolin P, editors. Amazonian Floodplain Forests: Ecophysiology, Biodiversity and Sustainable Management. New York: Springer Ecological Studies 210; 2010. p. 141-161. DOI 10.1007/978-90-481-8725-6_7

[35] Takahashi H, Yamauchi T, Colmer TD Nakazono M. Aerenchyma formation in plants. In: Van Dongen JT, Licausi F, editors. Low oxygen stress in plants, Plant Cell Monographs. 2014;**21**,247-265. DOI: 10.1007/978-3-7091-1254-0_13

[36] Piedade MTF, Junk W, D'Ângelo SA, Wittmann F, Schöngart J, Barbosa KMN, et al. Aquatic herbaceous plants of the Amazon floodplains: State of the art and research needed. Acta Limnologica Brasiliensis. 2010;**22**:165-178. DOI: 10.4322/actalb.02202006

[37] Striker GG, Colmer TD. Flooding tolerance of forage legumes. Journal of Experimental Botany. 2017;**68**:1851-1872. DOI: 10.1093/jxb/erw239

[38] Parolin P, De Simone O, Haase K, Waldhoff D, Rottenberger S, Kuhn U, et al. Central Amazonian floodplain forests: Tree adaptations in a pulsing system. The Botanical Review. 2004;**70**:357-380. DOI: 10.1663/0006-8101(2004)070[0357:CAFFTA]2.0.CO;2

[39] De Simone O, Müller E, Junk WJ, Schmidt W. Adaptations of Central Amazon tree species to prolonged flooding: Root morphology and leaf longevity. Plant Biology. 2002;**4**:515-522. DOI: 10.1055/s-2002-34127

[40] Seago JL. Anatomy of wetland plants. In: Finlayson CM, Everard M, Irvine K, McInnes RJ, Middleton BA, van Dam AA, Davidson NC, editors. Wetland book I: Structure and Function, Management, and Methods. Oswego: Springer; 2018. p. 363-374. DOI: org/10.1007/978-90-481-9659-3_45

[41] Pires AH. Anatomia dos órgãos vegetativos de *Genipa americana* L. e *Guazuma ulmifolia* Lam. ocorrentes em áreas alagáveis da Amazônia e em solos bem drenados do Cerrado [thesis]. Brasilia: Universidade de Brasília; 2020

[42] Somavilla NS, Graciano-Ribeiro D. Ontogeny and characterization of aerenchymatous tissues of Melastomataceae in the flooded and well-drained soils of a Neotropical savanna. Flora. 2012;**207**:212-222. DOI: 10.1016/j.flora.2012.01.007

[43] Herzog M, Pedersen O. Partial versus complete submergence: Snorkelling aids root aeration in *Rumex palustris* but not in *R. acetosa*. Plant Cell and Environment. 2014;**37**:2381-2390. DOI: 10.1111/pce.12284

[44] Watanabe K, Nishiuchi S, Kulichikhin K, Nakazono K. Does suberin accumulation in plant roots contribute to waterlogging tolerance? Frontiers in Plant Science. 2013;**4**:1-7. DOI: org/10.3389/fpls.2013.00178

[45] Yamauchi T, Abe F, Tsutsumi N, Nakazono M. Root cortex provides a venue for gas-space formation and is essential for pant adaptation to waterlogging. Frontiers in Plant Science. 2019;**10**:1-12. DOI: org/10.3389/fpls.2019.00259

[46] Batista CUN, Medri ME, Bianchini E, Medri C, Pimenta JA. Tolerância à inundação de *Cecropia pachystachya* Trec. (Cecropiaceae): aspectos ecofisiológicos e morfoanatômicos. Acta Botanica Brasilica. 2008;**22**:91-98 Available from: www.scielo.br/abb

[47] Ayup M, Chen Y, Nyongesa MJ, Zhang Y, Rajput V, Zhu Ch. Xylem anatomy and hydraulic traits of two co-occurring riparian desert plants. Iawa Journal.2015;**36**:69-83. DOI 10.1163/22941932-00000086

[48] Yang Ch, Zhanga X, Zhoua C, Seago Jr JL. Root and stem anatomy and histochemistry of four grasses from the Jianghan floodplain along the Yangtze river, China. Flora 2011;**206**:653-661. DOI: org/10.1016/j.flora.2010.11.011

[49] Pothasin P, Compton SG, Wangpakapattanawong P. Riparian *Ficus* tree communities: The distribution and abundance of riparian fig trees in northern Thailand. PLoS One. 2014;**9**:e108945. DOI: 10.1371/journal.pone.0108945

[50] James EK, Loureiro MF, Pott A, Pott VJ, Martins CM, Franco AA, et al. Flooding-tolerant legume symbioses from the Brazilian Pantanal. New Phytologist. 2001;**150**:723-738. DOI: 10.1046/j.1469-8137.2001.00126.x

[51] Martins CM, Borges WL, de Souza-costa J, Gouvêa-Rumjanek N. Rhizobial diversity from stem and root nodules of *Discolobium* and *Aeschynomene*. Acta Scientiarum, Agronomy. 2015;**37**:163-170. DOI: 10.4025/actasciagron.v37i2.19565

[52] Jackson MB. Plant survival in wet environments: Resilience and escape mediated by shoot systems. In: Bobbink R, Beltman B, Verhoeven JTA, Whigham DF, editors. Wetlands: Functioning, Biodiversity, Conservation and Restoration. Berlin: Springer-Verlag; 2006. pp. 16-36. DOI: 10.1007/978-3-540-33189-6

[53] Chen X, Visser EJW, de Kroon H, Pierik, Laurentius A. Voesenek CJ and Huber H. Fitness consequences of natural variation in flooding induced shoot elongation in *Rumex palustris*. New Phytologist. 2011;**190**:409-420. DOI: 10.1111/j.1469-8137.2010.03639

[54] Striker GG. Flooding stress on plants: Anatomical, morphological and physiological responses. Mworia JK, editor. Cap. 1. Botany, IntechOpen. 2012. p. 3-28. DOI: 10.5772/32922.

[55] Yamauchi T, Colmer TD, Pedersen O, Nakazono M. Regulation of root traits for internal aeration and tolerance to soil waterlogging-flooding stress. Plant Physiology. 2018;**176**:1118-1130. DOI: 10.1104/pp.17.01157

[56] Jackson MB, Ishizawa K, Ito O. Evolution and mechanisms of plant tolerance to flooding stress. Annals of Botany. 2009;**103**:137-142. DOI: 10.1093/aob/mcn242

[57] Kreuzwieser J, Rennenberg H. Molecular and physiological responses of trees to waterlogging stress. Plant, Cell and Environment. 2014;**37**:2245-2259. DOI: 10.1111/pce.12310

[58] Ruas EA, Ruas CF, Medri PS, Medri C, Medri CE, Bianchini E, et al. Anatomy and genetic diversity of two populations of *Schinus terebinthifolius* (Anacardiaceae) from the Tibagi river basin in Paraná, Brazil. Genetics and Molecular Research. 2011;**10**:526-536. DOI: 10.4238/vol10-1gmr1028

[59] Davanso VM, Medri ME, de Souza LA, Collis S. *Tabebuia avellanedae* Lor. Ex Griseb. (Bignoniaceae) submitted at the flooding ans the "ethrel" and silver nitrate application. Brazilian Archives of Biology and Technology. 2003;**46**:57-64. DOI: org/10.1590/S1516-89132003000100010

[60] Siegloch AM, Rodrigues-Dos Santos S, Cardoso JN. Marchiori. Estudo anatômico do lenho de *Colliguaya brasiliensis* Klotzsch ex Baill. (Euphorbiaceae). Balduinia. 2011;**27**:27-33. DOI: 10.5902/2358198014126

[61] Denardi, L. Anatomia eflexibilidade do caule de quatro espécies lenhosas para o manejo biotécnico de cursos de água [tese]. Santa Maria: Universidade Federal de Santa Maria; 2007.

[62] Raddatz DD, Machado SPF, DA Rosa RC, Sousa RDS, Sutili FJ, Marchiori CJN. Anatomia da madeira de *Aspidosperma riedelii* Müll. Arg. (Apocynaceae). Balduinia. 2018;**62**: 18-26. DOI: org/10.5902/2358198033223

[63] Mellerowicz EJ, Gorshkova TA. Tensional stress generation in gelatinous fibres: A review and possible mechanism based on cell-wall structure and composition. Journal of Experimental Botany. 2011;**63**:551-565. DOI: 10.1093/jxb/err339

[64] Cosmo NL, Kuniyoshi YS, Botosso PS. Anatomia da madeira de *Sebastiana commersoniana* (Baillon) Smith & Downs (Euphorbiaceae): aspectos funcionais e ecológicos. Acta Botanica Brasilica. 2010;**24**:747-755. Available from: www.scielo.br/abb e http://www.botanica.org.br/acta/ojs

[65] Piva TC, Rodrigues-Machado S, Scremin-Dias E. Anatomical and ultrastructural studies on gelatinous fibers in the organs of non-woody xerophytic and hydrophytic species. Botany. 2019;**97**:529-536. DOI: org/10.1139/cjb-2018-0220

[66] Monteiro de Oliveira D. Descrição e anatomia ecológica do xilema secundário de *Eugenia inundata* DC. (Myrtaceae) sob influência da inundação do lago Tefé, Amazonas, Brasil [tese]. Tefé: Universidade do Estado do Amazonas; 2019.

[67] Kröber W, Heklau H, Bruelheide H. Leaf morphology of 40 evergreen and deciduous broadleaved subtropical tree species and relationships to functional ecophysiological traits. Plant Biology. 2015;**17**:373-383. DOI: org/10.1111/plb.12250

[68] Parolin P. Seasonal changes of specific leaf mass and leaf size in trees of Amazonian floodplains. Phyton. 2002;**42**:169-185

[69] Waldhoff D, Parolin P. Morphology and anatomy of leaves. In: Junk W. Piedade M, Wittmann F, Schöngart J, Parolin P, editors. Amazonian Floodplain Forests: Ecophysiology, Biodiversity and Sustainable Management. Ecological Studies 2010;210: 179-202. DOI: 10.1007/978-90-481-8725-6_9

[70] Herrera A, Escala M, Rengifo E. Leaf anatomy changes related to physiological adaptations to flooding in Amazonian tree species.

Brazilian Journal of Plant Physiology. 2009;**21**:301-308. DOI: 10.1590/S1677-04202009000400006

[71] Alvarado G, García M, Jáuregui D, Alvarado H, Zambrano F. Leaf anatomy of six arboreal eudicotyledons species growing along an altitudinal gradient on the high basin of the Tocuyo river, Venezuela. Biologia. 2020;**75**:523-533. DOI: org/10.2478/s11756-020-00427-9

[72] Arambarri AM, Novoa MC, Bayón ND, Hernández MP, Colares MN, Monti C. Ecoanatomía foliar de árboles y arbustos de los distritos chaqueños occidental y serrano (Argentina). Boletin Sociedad Argentina de Botánica. 2011;**46**:1-21

[73] Crang R, Lyons-Sobaski S, Wise R. Plant Anatomy. A Concept-Based Approach to the Structure of Seed Plants. USA: Springer; 2019

[74] Jordan G, Dillon R, Weston P. Solar radiation as a factor in the evolution of scleromorphic leaf anatomy in Proteaceae. American Journal of Botany. 2005;**92**:789-796 Available from: https://www.jstor.org/stable/4126107

[75] Dezzeo N, Flores S, Zambrano S, Rodgers L, Ochoa E. Estructura y composición florística de bosques secos y sabanas en los llanos orientales del Orinoco, Venezuela. Interciencia. 2008;**33**:733-740 Available from: http://www.redalyc.org/articulo.oa?id=33931005

[76] Cabrera-Amaya DM, Rivera-Díaz O. Composición florística y estructura de los bosques ribereños de la cuenca baja del río Pauto, Casanare, Colombia. Caldasia. 2016;**38**:53-85. DOI: org/10.15446/caldasia.v38n1.57829

[77] Díaz WA, Rueda J, Acosta O, Martínez O, Castellanos YH. Composición florística del bosque ribereño del río San José, Reserva Forestal de Imataca, estado Bolívar, Venezuela. Acta Botanica Venezuelica. 2010;**33**:1-21 Available from: http://ve.scielo.org/scielo.php?script=sci_arttext&pid=S0084-59062010000100001&lng=es&nrm=iso

[78] Rangel-Ch JO. La biodiversidad de Colombia: significado y distribución regional. Revista Academia Colombiana de Ciencias Exactas Físicas y Naturales. 2015;**39**:176-200. DOI: org/10.18257/raccefyn.136

[79] Ballesteros-Correa J, Morelo-García L, Pérez-Torres J. Composición y estructura vegetal de fragmentos de bosque seco tropical en paisajes de ganadería extensiva bajo manejo silvopastoril y convencional en Córdoba, Colombia. Caldasia. 2019;**4**:224-234. DOI: org/10.15446/caldasia.v41n1.71320

[80] Prado E, Demarco D. Laticifers and secretory ducts: Similarities and differences. In: Hufnagel L, editor. Ecosystem Services and Global Ecology. Intechopen Cap. 6. 2018. pp. 103-124. DOI: 10.5772/intechopen.7570

[81] Ordoñez JC, Van Bodegom PM, Witte JM, Wright IJ, Reich PB, Aerts R. A global study of relationships between leaf traits, climate and soil measures of nutrient fertility. Global Ecology and Biogeography. 2009;**18**:137-149. DOI: org/10.1111/j.1466-8238.2008.00441.x

[82] Wright IA, Reich PB, Westoby M, Ackerly DD, Baruch Z, Bongers F, et al. The worldwide leaf economics spectrum. Nature. 2004;**428**:821-827. DOI: 10.1038/nature02403

[83] Cornwell WK, Ackerly DD. Community assembly and shifts in plant trait distributions across an environmental gradient in coastal California. Ecological Monographs. 2009;**79**:109-126

[84] Kajba D, Ballian D, Idžojtić M, Poljak I. Leaf morphology variation of

Populus nigra L. in natural populations along the rivers in Croatia and Bosnia and Herzegovina. South-east Eurean Forestry. 2015;**6**:39-51. DOI: 10.15177/seefor.15-06

[85] Pearce DW, Millard S, Bray D, Stewart F, Rood B. Stomatal characteristics of riparian poplar species in a semi-arid environment. Tree Physiology. 2005;**26**:211-218 Available from: https://academic.oup.com/treephys/article-abstract/26/2/211/1676809

[86] Hetherington AM, Woodward FI. The role of stomata in sensing and driving environmental change. Nature. 2003;**424**:901-908. DOI: 10.1038/nature01843

[87] Rood SB, Mahoney JM. River damming and riparian cottonwoods along the Marias River, Montana. Rivers. 1995;**5**:195-207

[88] Rood SB, Braatne JH, Hughes FMR. Ecophysiology of riparian cottonwoods: Streamflow dependence, water relations and restoration. Tree Physiology. 2003;**23**:1113-1124. DOI: 10.1093/treephys/23.16.1113

[89] Čortan D, Vilotić D, Šijačić-Nikolić M, Miljković D. Leaf stomatal traits variation within and among black poplar native populations in Serbia. Bosque. 2017;**38**:337-345. DOI: 10.4067/S0717-92002017000200011

[90] Grašic M, Pibercnik M, Zelnik I, Abram D, Gaberšcik A. Invasive alien vines affect leaf traits of riparian woody vegetation. Water. 2019;**11**:2395. DOI: 10.3390/w11112395

[91] Parolin P, Waldhoff D, Piedade MTF. Fruit and seed chemistry, biomass and dispersal. In: Junk WJ, Piedade MTF, Wittmann F, Schöngart J, Parolin P, editors. Amazonian Floodplain Forests: Ecophysiology, Biodiversity and Sustainable Management. New York: Springer Ecological Studies. 2010;210. p. 243-258. DOI 10.1007/978-90-481-8725-6_11

[92] Nilsson C, Brown RL, Jansson R, Merritt DM. The role of hydrochory in structuring riparian and wetland vegetation. Biological Reviews. 2010;**85**:837-858. DOI: 10.1111/j.1469-185x.2010.00129.x

[93] Horn MH, Correa SB, Parolin P, Pollux BJA, Anderson JT, Lucas C, et al. Seed dispersal by fishes in tropical and temperate fresh waters: The growing evidence. Acta Oecologica. 2011;**37**:561-577. DOI: 10.1016/j.actao.2011.06.004

[94] Myers RL. Remoción y dispersión de semillas en un humedal dominado por palmas (Arecaceae) en el noreste de Costa Rica. Revista Biologia Tropical. 2013;**61**:67-85 http://www.redalyc.org/articulo.oa?id=44958811006

[95] Kubitzki K, Ziburski A. Seed dispersal in flood plain forests of Amazonia. Biotropica. 1994;**26**:30-43. DOI: 10.2307/2389108

[96] Scarano FR, Pereira TS, Rôças G. Seed germination during floatation and seedling growth of *Carapa guianensis*, a tree from flood-prone forests of the Amazon. Plant Ecology. 2003;**168**:291-296. DOI: org/10.1023/A:1024486715690

[97] Riis T, Sand-Jensen K. Dispersal of plant fragments in small streams. Freshwater Biology. 2006;**51**:274-286. DOI: 10.1111/j.1365-2427.2005.01496.x

[98] Worbes M. The forest ecosystem of the floodplains. In: Junk WJ, editor. The Central Amazon Floodplain Ecological Studies, Vol. 126. Berlin Heidelberg: Springer; 1997. p. 223-265. DOI: org/10.1007/978-3-662-03416-3_11

Chapter 8

Temporary De-Poldering for a Long Term Flood/Sediment Management in the Southwestern Bangladesh

Rocky Talchabhadel, Kenji Kawaike and Hajime Nakagawa

Abstract

Southwestern Bangladesh has been seriously affected by perennial waterlogging over the last few decades. It is primarily due to excessive riverbed siltation outside the polders after the construction of embankments along both sides of the tidal rivers. These embankments de-linked the huge natural floodplains and restricted a gradual process of natural deposition inside the polders. An introduction of the tidal basin concept by temporary de-poldering (embankment cut) at some designated locations has substantially solved the issues. The current chapter looks at the historical practice of flood/sediment management, the evolution of embankments and their de-poldering, inclusion of Tidal River Management (TRM) in long term flood/sediment management, and discusses a technical aspect of flood/sediment dynamics across the tidal river system. The process of restoring beneficial tidal flooding by cutting embankment at certain locations, commonly known as TRM, is not a novel method. The TRM has started from age-old practice and proves technically one of the effective methods of sustainable flood/sediment management in the tide-dominated river system. It is an example of building with nature, where little human interventions are needed, and a resilient measure for waterlogging, drainage-congestion, and river-siltation.

Keywords: *beel*, de-polder, polder, siltation, Tidal River Management (TRM), waterlogging

1. Introduction

For millennia the Bengal delta has been home to dynamic interplay of sediment, water, and land. The southwestern region of Bangladesh, comprised mainly of the Khulna, Jessore, Satkhira, and Bagerghat (shown in **Figure 1**), is characterized by an agro-economical system with numerous morphologically active tidal rivers and brackish water regime. Before the lowlands were enclosed by earthen embankments, that keep the main tidal channels outside the polder, the tidal river would inundate vast tract of lowland two times in a day. In addition, there were less population pressure and lower level of agricultural production. In the past *Zamindars* (landlords) were managing the water resources in an open-wetland ecosystem of

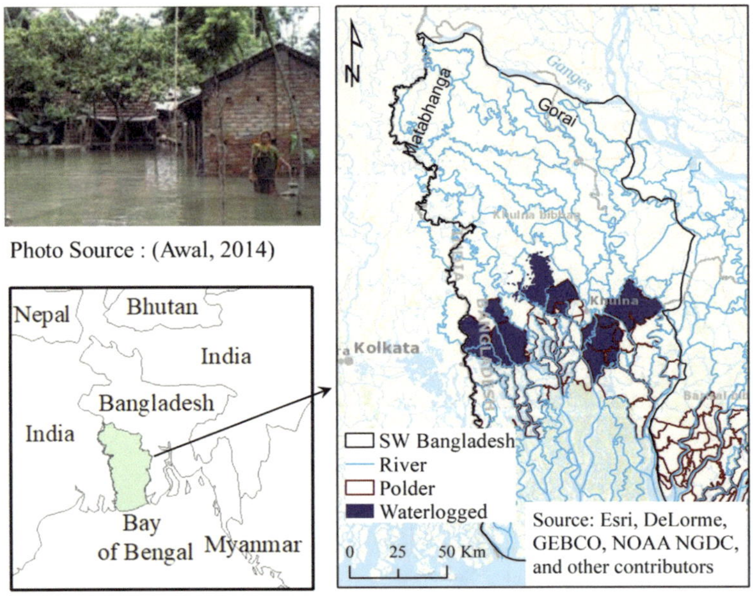

Figure 1.
Map showing prevailing waterlogging in the coastal polders across the southwestern (SW) region of Bangladesh. A representative photo [1] shows permanent waterlogging in Tala, Satkhira in august 2011.

southwestern delta. They used low earthen embankments during eight dry months of the year (commonly known as "*ostomasi bundh*" in local *Bengali* language) for prevention of tidal intrusion and protection of agricultural lands. After the harvest, the embankments were dismantled or swept away by natural river flood during monsoon [2]. The flood water during the monsoon enabled the deposition of sediment, nutrients in the form of over land irrigation, and soil nourishment and also allowed for seasonal fishing [3].

After the abolition of *Zamindari* system (land holding by *Zamindars*), the maintenance became disrupted. In addition, there were disastrous floods in 1954, 1955 and 1956 which caused large-scale damage to human lives and food crop [4]. In order to meet the food demand of rapidly increasing population there was a need of all-round the year cultivation which requires a proper flood control [5]. In the 1960s, the Coastal Embankment Project (CEP) constructed 123 polders in southwestern coastal region of Bangladesh. Polders were constructed to protect land and livelihood from floods, salinity intrusion and to facilitate increased agricultural production. Daily tidal intrusions became a thing of past [6] and with two-three harvests annually the agriculture-based economy gained significantly for 10–15 years.

At the same time, coastal embankments denied the entry of the tides into the polders and gradually prevented the slow process of land formation on flood plains. As a result, unintended heavy loads of riverbed siltation [7] altered the natural river flow seriously and tidal rivers lost their navigability. Several rivers completely dried within a few years to few decades. The floodplains inside the polders gradually lowered than riverbed outside the polders and closed the exits of the sluiced gates. Adding to the problem, the Farakka Barrage across the Ganges River reduced the freshwater flow since 1976. In the southwestern region of Bangladesh, the Matabhanga and Gorai Rivers (shown in **Figure 1**) are the main distributaries of the Padma River for freshwater upland flow. Moreover, due to the heavy siltation at the rivers' mouth, their distributaries virtually gets no water from upstream in dry season. These tidal rivers could not effectively drain the nearby *beel*s (low lying land or natural depression) inside polders anymore, which results in serious waterlogging (shown in a representative photo in **Figure 1**). There is water everywhere but not drinking water.

2. Emergence of de-poldering

Since the 1980s, vast tracts of land went under water semi-permanently; the rainwater in the region could not drain resulting to a long-standing waterlogging problem (for more than 6 months in a year). Major flood events in 1987, and 1988 triggered large-scale long-standing waterlogging problems. In an apparent move to address waterlogging especially in the *beel Dakatia*, the largest *beel* in southwestern region of Bangladesh (shown in **Figure 2**), the Khulna Coastal Embankment Rehabilitation Project (KCERP) was introduced in the late 80s with support from Asian Development Bank (ADB). There was lack of consultation with local concerned people and their needs were hardly reflected in the project design. The waterlogging problem further deteriorated leading to widespread public protest [5].

The project was ultimately suspended as local people cut the embankment at four places during a *mahashamabesh* (mass community mobilization) with the intention of draining away water from the *beel* in September 1990. They believed the tidal flow could be restored through breaching, the land would rise through the accumulation of silt and the stored water coming out could push through the narrow river [4]. The restored open connection with the river immediately initiated

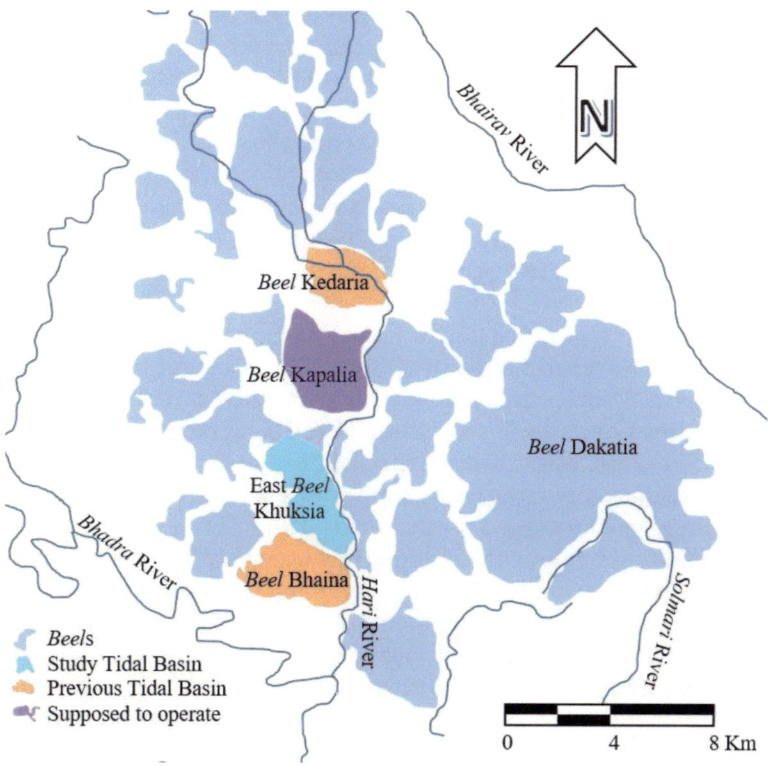

Figure 2.
Location of Beel *Dakatia,* beel *Kedaria,* beel *Kapalia, east* beel *Khuksia, and* beel *Bhaina in Hari-Mukteswari River system. Note: Other* beels *(unnamed) are visible in shape and locations (adapted from Talchabhadel et al. [8]).*

tidal dynamics and sedimentation processes within the beel. Approx. 1000 ha of high land had been gained in the *beel* Dakatia via public cuts (i.e. without any technical studies on design of de-poldering).

To solve drainage congestion issue, enhance flooding protection (both tidal and seasonal), and rehabilitate existing drainage infrastructure the Khulna-Jessore Drainage and Rehabilitation Project (KJDRP) started in 1993. For Bangladesh Water Development Board (BWDB), the public cutting of embankments was unacceptable because it went against the law. The cuts were closed in *beel* Dakatia by BWDB in 1994; the land level was elevated substantially. The *beel* Dakatia public cutting enhanced the tidal free flow and restored the navigability of the river by increasing both depth and width within a short span of time. But then, the river largely died when the cut points were filled up.

The river was almost rescued by local people through their intuitive knowledge but ultimately it died. The dry river bed is covered by paddy field and homesteads. Although international consultants had incorporated plans for controlled tidal flooding in the early 1990s, BWDB did not include the practice in the KJDRP, arguing that it misrepresented the real problem and that it was not scientifically grounded [3]. The waterlogging scenario in southwestern region was intensified in 1997 flood (monsoon) season. The water could not be drained away overland, nor could it be discharged. In post flood season, local people had cut the embankment of *beel* Bhaina for quicker drainage needed for Boro-rice cultivation on their land. Although rapid drainage and recession of water occurred from *beel* Bhaina due to high magnitude of head difference the *beel* could not be brought under Boro-rice cultivation by closing the cut-point. It became gradually wider and deeper and went beyond the capacity of the local people to close. But at the end of the dry season, they observed that land level of *beel* Bhaina was raised, on the other hand, depth of Hari River increased significantly. The temporary restoration of tidal flooding by de-poldering came to be known conceptually as TRM. The basic idea in this process is to allow high tides bringing muddy water flow with a thick concentration of sediment into designated tidal basin and releasing the tidal flow back into the river during the low tides. During the repetitive process of such tidal restoration, a large chunk of sediment deposits on the tidal basin which otherwise would deposit in the river channel if there were no de-poldering. In addition, during low tides the flow even washes away sediment from river bed making the river congestion free and more navigable.

It brought the attention of the donor agency to apply these phenomena in all the *beel*s adjacent to Hari River sequentially as a tool to remove waterlogging from the inundated *beel*s. A socio-environmental impact study carried out by research institute EGIS (now Center for Environmental and Geographic Information System - CEGIS) in 1998 was crucial in the debate of open and closed approaches of flood/sediment management: the study supported closed and embanked approach (i.e. polder), but it equally argued for the inclusion of a TRM (de-polder) [9].

An iconic public embankment cutting in *beel* Dakatia proved to be the centerpiece of the discussion on open (de-polder and temporary water retention) versus closed (embanked and detached catchment) approaches. With *beel* Dakatia and *beel* Bhaina together, there was a decade of local people driven TRM. The attention to environmental issues and an emphasis on social participation have (forcefully) created some space for de-polder in the domain of embanked flood management. Since the *beel* Bhaina was taken up by KJDRP, it is considered as the first TRM officially. In 2002, BWDB incorporated TRM in *beel* Kedaria, then in east *beel* Khuksia in 2006. It showed that, over time, a dominant paradigm suggesting poldering was supplemented by some incidental occasions of de-poldering. The description of operated TRMs are summarized in **Table 1**.

Technical Context	Informal - Local people proactive		Formal	
	Beel Dakatia	*Beel* Bhaina	*Beel* Kedaria	East *Beel* Khuksia
	11000 ha	600 ha	600 ha	800 ha
	(1990–1994)	**(1997–2001)**	**(2002–2005)**	**(2006–2012)**
Strategy	No technical planning	No technical planning, started by local people and undertaken by BWDB later	Planned and approved by BWDB	Planned and approved by BWDB
Embankment Opening	Public Cut in 4 places	Public Cut and continuous operation	No breaching of embankment, operated through 21 vent Bhabodah regulator (6 months per year)	Formal De-poldering and continuous operation
Peripheral Embankment	No	No	Yes	Yes
Land Heightening	Not measured exactly; Around 10% of such a large area significantly land heightened	Around 60% uniformly silted; Average of 0.8 m land heightening (1.5–2.0 m at cut point and 0.2 m in the far end); Net silt deposition = 6.48 million m^3	Insignificant Siltation; Net silt deposition = 0.5 million m^3	Around 50% uniformly silted; Average of 1.2 meter land heightening (1.5–2.0 m at cut point and 0.5 m in the far end)
Compensation	No	No	No	Yes (not easy to get)
Stakeholder Participation	Local Communities and NGOs, Social and political activists	Local Communities and NGOs, Social and political activists	Local Communities and NGOs, Social and political activists, Government institutions, research organizations, donors and so on	Local Communities and NGOs, Social and political activists, Government institutions, research organizations, donors and so on
Conflicts	Local conflicts were resolved		Conflicts exist among local stakeholders (shrimp-field owners, local farmers, landless people, daily-wage workers, NGOs, civil society organizations) as well as between local stakeholders and government institutions (local government, Union Parishad, Upazilla, district administrative, BWDB,)	

Table 1.
Summary of description of operated TRMs (both informal and formal) [2, 4, 10, 11].

3. Shifting of Tidal River Management

The people of *beel* Kedaria were very much in favor of TRM to raise their land, but it did not meet their expectation. Although there was no waterlogging in the area during the operation the landowners lost their interest to operate TRM by submerging their land without any yield and compensation for three years. This created a negative impact on TRM and ultimately BWDB could not find any *beels* to operate TRM in the year 2005. The riverbed silted up immediately after stopping the operation of TRM shown in **Figure 3**. After the termination of TRM, the riverbed rose almost by 4–6 meters due to reduced tidal prism. It clearly indicates the area will remain inundated if TRM is not operated in any of the *beel*s. To solve the problems in the long run, TRM operations are to be shifted/rotated effectively in different *beel*s. It is reported that the shifting of TRMs from downstream to upstream yields larger sediment deposition than from shifting TRMs from upstream to downstream [6].

The challenges of operation are social and institutional. People are unwilling to provide their land since they cannot do economic activities like agriculture or fisheries. To overcome this challenge a compensation mechanism for crop and fisheries has been established then after. During the operation of TRM in east *beel* Khuksia, compensation mechanism was developed.

The plan of shifting of TRMs has been included in long-term policy for the southwestern region. Planned large-scale rotational TRMs were then developed in major river systems. Such as Hari-Mukteswari river system: East Khuksia, Kapalia, Baruna, Payra, West Khuksia and Singha by 2047 [12]; Kobadak river system: Pakhimara, Hariharnagar, Raruli, Rajapur Harinkhola, Delua and Jalalpur by 2045 [13].

In Hari-Mukteswari river system, *beel* Kapalia was planned to be operated as TRM from 2012 onwards for a period of 5 years, but violent protests prevented this. Some of the major reasons are compensation mechanism and lack of trust in government. East *beel* Khuksia was operated for 7 years but the initial plan was to only operate for 4 years. BWDB is currently bringing back the planned operation of TRM in Hari-Mukteswari river system (including *beel* Kedaria) by addressing pertinent social, economic, technical, and political problems. Meanwhile, BWDB is replicating the TRM concept in other receptive river system such as Kobadak river system. *Beel* Pakhimara is at present operational since 2015.

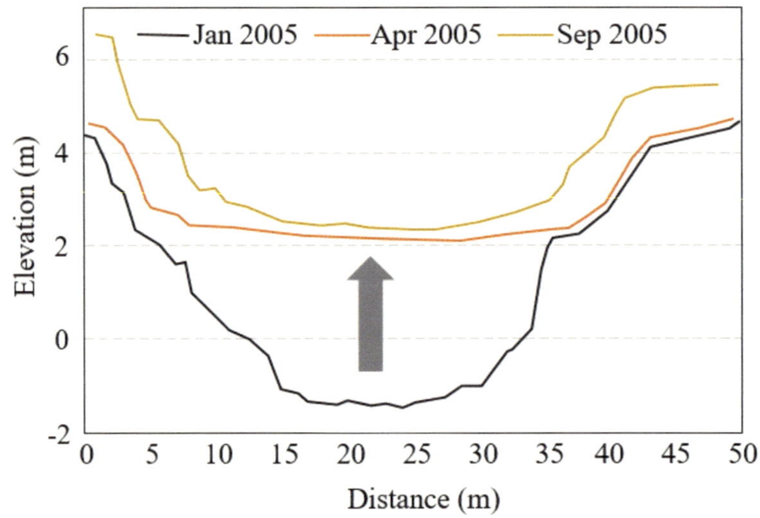

Figure 3.
Impact on Hari River morphology at Ranai after termination of Tidal River Management operation in beel *Kedaria (source: IWM [12]).*

4. Tidal River Management for flood/sediment management

Dredging and excavation are costly and are not eco-friendly. They are again more time consuming to manage huge drainage congestion. In addition, the deposited sediments by excavating again fall into the river through runoff in a rainy season. De-poldering and then controlled flooding (during TRM process) in a particular *beel* is not a new way of sediment management. However, the method involves taking full advantages of the natural tide movement with little human interventions. This concept involves temporary de-poldering to allow *jowar-bhata khelano* (free play of tidal flow) in selected *beel* through a link canal shown in **Figure 4a**. At high tide, muddy water enters with a thick concentration of sediment and deposits large chunks of sediment on the selected *beel*. At low tide, relatively clearer water erodes the river bed, increasing the depth of the river. This process is a participatory approach where the people of the identified tidal basin have to provide their land remain flooded for an intended period (about 5 years depending on tidal prism and area of the *beel*). **Figures 4a–c** and 5 show that TRM reclaims land from waterlogging and inundation, increases the river navigability by increasing river depth and width, and provides solution of drainage congestion of southwestern region.

This concept is technically feasible, environmentally friendly and socially acceptable [11, 15]. The experiences of TRM operated so far, exhibited that there was no

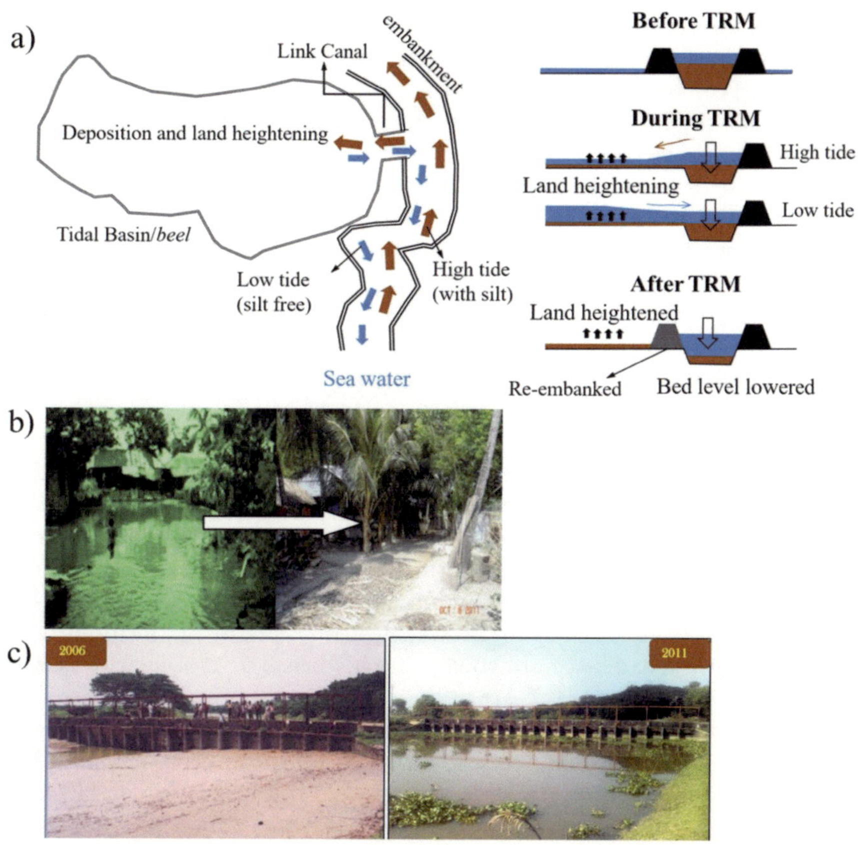

Figure 4.
(a) Tidal basin concept: Free play of tidal flow in selected beel, *(b) land reclamation in Arua village in between 2005 and 2011 [11], and (c) revival of Hari River near Bhabodah regulator in between 2006 and 2011 [14] (adapted from Talchabhadel et al., [8]).*

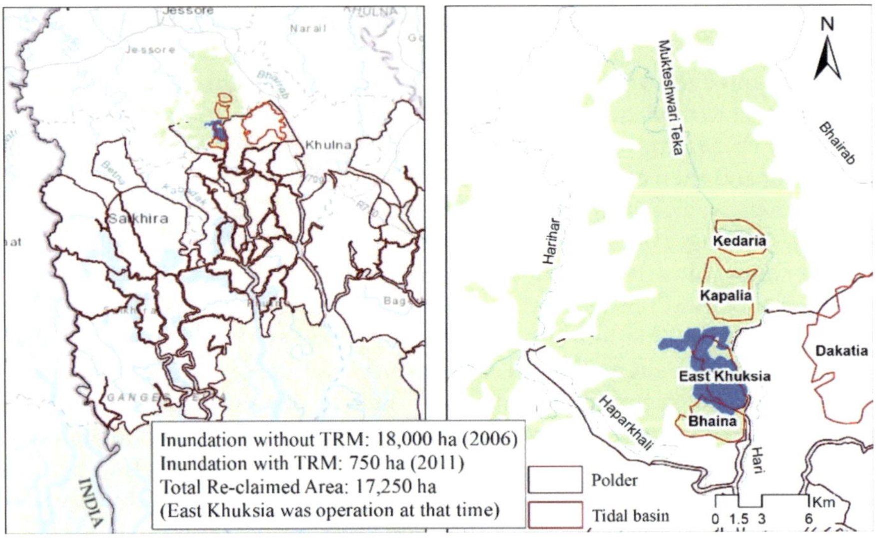

Figure 5.
Comparison of inundation area with and without operation of TRM (adapted from Talchabhadel et al., [8]).

drainage congestion and waterlogging problem throughout in the river system and the selected *beel*s were heightened significantly. Though the TRM operated *beel* is not suitable for practicing any agriculture, about 20–25 surrounding *beels* located upstream could have favorable ecological conditions for better agricultural production. And, TRM can increase drainage capacity to a long distance up to 20 km of the lower stream at a single intervention from the place of TRM implementation [16].

Several studies have been conducted qualitative/quantitative analyses of TRM and its relation to disaster, ecosystem, environment, agriculture and suggested the appropriate design of TRM. Some studies [17–19] have analyzed the effectiveness with different interventions like crossing dam, dredging, and compartmentalization. Some [20–22] have suggested optimum opening sizes of link canals connecting the tidal basin and tidal river. The TRM is effective when there is no upland flow or the upland flow is reduced, therefore, crossing dam at upstream of cut point is recommended during the operation of TRM. There have been still issues related to non-uniform sediment deposition across the selected *beel* despite of several studies based on experimental and numerical simulation. Higher sedimentation occurs close to the opening and decreases gradually to the further end of the *beel* [23]. To solve this issue, one option could be construction of embankment on both sides of main channel inside tidal basin and allowing siltation from most remote areas and cutting embankment part by part from remote end to link canal or introduce compartmentalization [18]. Also, the alignment of link canal controls the area of high shear stress [22]. A relationship between the tidal prism and minimum area for a tidal equilibrium can be established using historical data of nearby tidal rivers while designing the link canal [17, 21].

A combination of TRM with some degree of dredging/excavation could serve effective drainage and sediment management. Most importantly, the operation of TRM should be continuous. In the absence of sequential or shifting of TRMs, the acquired benefits might be wasted due to repetitive heavy siltation. Therefore, there should not be any time gaps between closing of one *beel* and operation of successive TRM *beel* [24]. Effective operation of TRMs could contribute to flood alleviation in southwestern Bangladesh and more than 100 beels are suitable of operation of TRM process [7]. However, during the operation of TRM agriculture is affected for a few years in operated *beel*. A systematic program of TRM is therefore necessary in order to increase overall agricultural production by reducing flood susceptibility and managing sediments. A manageable size of the tidal basin at one intervention is around 1000–2000 ha.

5. Toward sustainable solution

TRM evolved into an environmentally accepted flood/sediment management practice based on indigenous knowledge and it contributes to land heightening, flood resistance, and food security. However, the socio-political use of TRM complicates its implementation and there have been well-documented failures of TRM implementation. The effectiveness of TRM has not matched the expectation of majority of stakeholders. The challenges include engineering, socio-political, and institutional. Overall, a key barrier to the successful implementation is a proper management of compensation. Though the compensation mechanism started in east *beel* Khuksia, there were conflicts among different stakeholders (farmers, landless people, daily-wage workers, shrimp-field owners, NGOs, local organizations) and between these stakeholders and government bodies (BWDB, local government, district administrative) are prevalent [6]. The conflicts are related to compensation amount, procedure of compensation, land use practices, employment, shifting of TRMs and overall coordination.

Generally, the landless farmers and poor fishermen are the main sufferers of TRM implementation. Because they worked under landowners and they would not get the land compensation [25]. It is a high time to identify programs addressing their needs, for instance, provision for the cooperative fish culture. Shrimp farmers, who did not want to lose their leased land under any circumstances are against the TRM [4].

BWDB officials being primarily engineers and hydrologists, they may lack the skills and expertise to facilitate stakeholder [26]. The river channels should be kept open as far as possible. Unplanned drainage obstacles like *gher*s (shrimp aquaculture ponds) should be removed and appropriate policy should be developed to stop misuse of *khal*s (drainage canals). Sediment management for uniform distribution is required to improve TRM practices. Highly saline zones are unsuitable for the TRM sites as high saline sediment precludes agriculture. A temporary cross dam should be constructed in the dry season. If possible, the river should be re-excavated before the monsoon and the crossing dam should be made open during the monsoon. Peripheral embankment should enclose the selected *beel*. A simple mechanism of distributing the crop compensation has to be established. There is a strong need of an open, transparent and inclusive planning, implementation, and operation-maintenance of TRM.

The evolution of social learning, and participation of different stakeholders clear indicate a need for an approach that considers multi-dimensional (social, economic and environmental) consequences [4]. Also, bringing together people from different backgrounds, perspectives, values does not automatically lead to effective management. It requires a multi-loop process in improving a multiple-actor-based management practice [2]. Key criteria for successful TRM implementation are:

- Intensive consultation with concerned communities, and affected landowners

- Collaborative working (conflict resolution among different land-use groups, working agencies, and government institutions)

- Strengthening of local institutions
- Regular supervision by government institutions
- Continuous information and motivational campaign
- Interdisciplinary research (technical and socio-economic research for proper selection of *beel*s, their operation, and compensation distribution)
- Iterative learning (evaluation of success and failure to guide for the effective operation in next *beel*)
- Sequential implementation of TRM (no stoppage of operation at any cost)

A long-term strategy of Bangladesh Delta Plan (up to 2100) has highly acknowledged the TRM concept for solving waterlogging problems and making river alive. Restoration of tidal plain allowing tidal inundation is an effective measure for increasing the tidal prism, raising the low lying land (sediment-starved flood plain), and ultimately solving waterlogging issues with sustainable sediment management. In order to obtain faster drainage, pumping might be required in addition to gravity drainage of polders. The iterative learning in coming days will surely move toward the sustainable solutions. Although an embankment along the coast is required, sediment management is also essential for sustainability. The concept of TRM, evolved from indigenous knowledge and blended with scientific understanding, should go hand in hand with other water/sediment-related interventions.

6. Conclusion

Unlike excavating, TRM is eco-friendly, cost-effective (no need of large-scale engineering), and time-efficient (maximum utilization of natural tidal flow). It simply allows beneficial flood in raising low-lying land, scouring silted river, and improving drainage congestion. The TRM process restores the tidal prism and provides sufficient space for the river. However, internal roads, cultivation, and economic activities are hampered during the TRM operation period. The TRM concept is one of the effective measures of managing flood/sediment problems in tide-dominated river system.

Evolved from traditional wisdom, the TRM process is blended with scientific knowledge and it proves to be a sustainable way forward for effective flood/sediment management. However, a key barrier to the successful implementation is compensation. The landless farmers and poor fishermen are the main sufferers of TRM implementation. Programs focused for main sufferers are equally needed along with a simple mechanism of distribution of the compensation to land owners.

Since the TRM is a participatory approach, all the organizations and stakeholders connected with the different stages of the development process of TRM should be responsible for promoting the entire process. Multiple-stakeholders-based inclusive planning, operation-management practice with iterative learnings are necessary for successful implementation of TRM concept. With time we need to rethink and recalibrate the roles of stakeholders in order to manage possible conflicts and maintain impulsive monitoring and evaluation of all groups involved from local communities to government bodies. Therefore, the process need continuous updating based on the monitoring results.

Acknowledgements

The research was supported by JST/JICA SATREPS program on disaster prevention/mitigation measures against floods and storm surges in Bangladesh (PI: Dr. Hajime Nakagawa). The first author was supported by a Monbukagakusho (MEXT) scholarship during his Ph.D. (2014 – 2017) and the Japan Society for the Promotion of Sciences (JSPS) Postdoctoral Fellowship Program, 2019 – 2020, (grant in aid P19052; Host: Dr. Kenji Kawaike).

Conflict of interest

No conflict of interest.

Author details

Rocky Talchabhadel[1]*, Kenji Kawaike[2] and Hajime Nakagawa[2]

1 Texas A&M AgriLife Research, Texas A&M University, El Paso, Texas, USA

2 Disaster Prevention Research Institute, Kyoto University, Kyoto, Japan

*Address all correspondence to: rocky.ioe@gmail.com

References

[1] Awal MA. Water logging in south-western coastal region of Bangladesh: local adaptation and policy options. Sci Postprint. 2014;1(1):1-13.

[2] Mutahara M, Warner JF, Wals AEJ, Khan MSA, Wester P. Social learning for adaptive delta management: Tidal River Management in the Bangladesh Delta. Int J Water Resour Dev. 2018;34(6):923-943.

[3] Staveren MF Van, Warner JF, Khan MSA. Bringing in the tides. From closing down to opening up delta polders via Tidal River Management in the southwest delta of Bangladesh. Water Policy. 2017;19:147-164.

[4] Gain AK, Benson D, Rahman R, Datta DK, Rouillard JJ. Tidal river management in the south west Ganges-Brahmaputra delta in Bangladesh: Moving towards a transdisciplinary approach? Environ Sci Policy. 2017;75(May):111-120.

[5] Haque KNH, Chowdhury FA, Khatun KR. Participatory environmental governance and climate change adaptation: mainstreaming of tidal river management in south-west Bangladesh. In: Ha H, editor. Land and Disaster Management Strategies in Asia. 2015. p. 189-208.

[6] Talchabhadel R, Nakagawa H, Kawaike K. Selection of Appropriate Shifting of Tidal River Management. In: Water, Flood Management and Water Security Under a Changing Climate. Cham: Springer International Publishing; 2020. p. 283-299.

[7] Adnan MSG, Talchabhadel R, Nakagawa H, Hall JW. The potential of Tidal River Management for flood alleviation in South Western Bangladesh. Sci Total Environ. 2020 Aug;731:138747.

[8] Talchabhadel R, Nakagawa H, Kawaike K, Sadik MS. Polder to De-polder: an Innovative Sediment Management in Tidal Basin in the Southwestern Bangladesh. Annu Disaster Prev Res Inst. 2018 Feb;(61B):623-630.

[9] EGIS. Environmental and Social Impact Assessment of the Khulna-Jessore Drainage Rehabilitation Project, BWDB. Dhaka, Bangladesh; 1998.

[10] Leendert Ferdinand de D. Tidal River Management Temporary depoldering to mitigate drainage congestion in the southwest delta of Bangladesh. Wageningen University; 2013.

[11] Rezaie AM, Islam T, Rouf T. Limitations of institutional management and socio-economic barriers of Tidal River Management, a semi-natural process to save bhabodaho from water-logging problem. Fukuoka S, Nakagawa H, Sumi T, Zhang H, editors. Adv River Sediment Res. 2013;2173-2183.

[12] IWM. Feasibility Study and Detailed Engineering Design for Long Term Solution of Drainage Problems in the Bhabodah Area. Dhaka, Bangladesh; 2010.

[13] IWM. Feasibility Study for Sustainable Drainage and Flood Management of Kobadak River Basin under Jessore and Satkhira Districts. 2010.

[14] Paul A, Nath B, Abbas MR. Tidal River Management (TRM) and its implication in disaster management : A geospatial study on Hari-Teka river basin, Jessore,. Int J Geomatics Geosci. 2013;4(1):125-135.

[15] Rezaie AM, Naveram UK. Tidal river management : An innovative approach

for terminating drainage congestion and raising land through sedimentation in the Bhabodaho area, Bangladesh. Fukuoka S, Nakagawa H, Sumi T, Zhang H, editors. Adv River Sediment Res. 2013;1363-75.

[16] Al M, Naher N, Azadi H, Passel S Van. Sustainability impacts of tidal river management : Towards a conceptual framework. Ecol Indic [Internet]. 2018;85(November 2017):451-67. Available from: https://doi.org/10.1016/j.ecolind.2017.10.022

[17] Shampa, Pramanik MIM. Tidal River Management (TRM) for Selected Coastal Area of Bangladesh to Mitigate Drainage Congestion. Int J Sci Technol Res. 2012;1(5):1-6.

[18] Ibne Amir MSI, Khan MSA, Kamal Khan MM, Golam Rasul M, Akram F. Tidal river sediment management - a case study in southwestern Bangladesh. Int J Civil, Archit Struct Constr Eng. 2013;7(3):183-193.

[19] Rahman MZ, Islam MS, Khan ZH. Tidal River Management (TRM)-An Innovative Scientific Approach for Sustainable Sediment Management. In: international Conference on Recent Innovation in Civil Engineering for Sustainable Development (IICSD-2015). 2015. p. 954-959.

[20] Stive MJF, Rakhorst RD. Review of empirical relationships between inlet cross-section and tidal prism. J Water Resour Environ Eng. 2008;23(23):89-95.

[21] Talchabhadel R, Nakagawa H, Kawaike K, Hashimoto M, Sahboun N. Experimental investigation on opening size of tidal basin management : A case study in southwestern Bangladesh. J Japan Soc Civ Eng Ser B1 Hydraulic Eng. 2017;73(4):I_781-I_786.

[22] Talchabhadel R, Ota K, Nakagawa H, Kawaike K. Three-dimensional simulation of flow and sediment transport processes in tidal basin. J Japanese Soc Civ Eng Ser B1 Hydraulic Eng. 2018;74(4):I_955-I_960.

[23] Talchabhadel R, Nakagawa H, Kawaike K. Sediment management in tidal river : A case study of East Beel Khuksia, Bangladesh. In: RiverFlow 2018. 2018. p. 1-10.

[24] Ullah MW, Mahmud S. Appropriate planning, design and implementation modalities for successful application of Tidal River Management (TRM) in coastal delta. In: 6th International Conference on Flood Management (ICWFM-2017). 2017. p. 133-140.

[25] Talchabhadel R, Nakagawa H, Kawaike K. Tidal River Management (TRM) and Tidal Basin Management (TBM): A case study on Bangladesh. Lang M, Klijn F, Samuels P, editors. E3S Web Conf. 2016 Oct 20;7:12009.

[26] Gain AK, Schwab M. An assessment of water governance trends: The case of Bangladesh. Water Policy. 2012;14(5):821-840.

Chapter 9

Agricultural Activities and Restoration of Lake Chad

Patrick Arnold Ombiono Kitoto

Abstract

The disappearance of 90% of Lake Chad's surface has brought riparian countries to elaborate a restoration project for this natural asset. The aim of this study is to estimate the benefits and costs associated with the realization of this project, in order to determine if it is socially profitable. The methodological approach use data from the contingent valuation survey conducted in 2011 in the Cameroonian part of Lake Chad and appropriate statistical and econometric procedures. First, we estimate the middle and long term benefits of the project to be €5,549,576.832 and €38,543,518.56 respectively. Then, we evaluate the costs generated by the implementation of such a project to €37,960,149.12. Finally, the social profitability of this project depends on the temporal horizon used by decision-makers. It is negative for an economic horizon and positive for an ecological horizon.

Keywords: Lake Chad, costs-benefits analysis, contingent valuation method, social benefits, social costs, temporal horizon

1. Introduction

Situated at the heart of the African continent, in an arid Sahelian environment where every water point is a source of life, Lake Chad is at the heart of a complex issue related to the choice between economic activities and ecological balance. For a long time, priority was given to economic activities which, combined with the growing effects of global warming, have particularly led to the loss of 90% of its water surface area, thus going from 25,000 Km^2 (in 1964) to less than 2,500 Km^2 today and exposing the local resident populations that depend on it to climatic vulnerabilities [1]. NASA mapping surveys in 2001 (**Figure 1-A below**), corroborated by satellite imagery (**Figure 1-B** below), confirm this. These figures also allow us to note that the northern part of Lake Chad, integrating Niger and Nigeria, has completely disappeared, and is taken over either by vegetation or expanses of sand, or by human settlements.

Today, safeguarding Lake Chad is a major concern for the planet. This concern is reflected in the willingness of the riparian states to restore this natural asset, through the construction of an inland waterway through which it would be supplied with water. In addition to its very high cost of construction and despite a high proportion of favorable opinions, this project is not socially desirable in the sense of Pareto since it could generate another cost for society. Indeed, the restoration of Lake Chad is likely to have contrasting effects not only on the environment, but also on the population engaged in agricultural activities on its shores and surroundings.

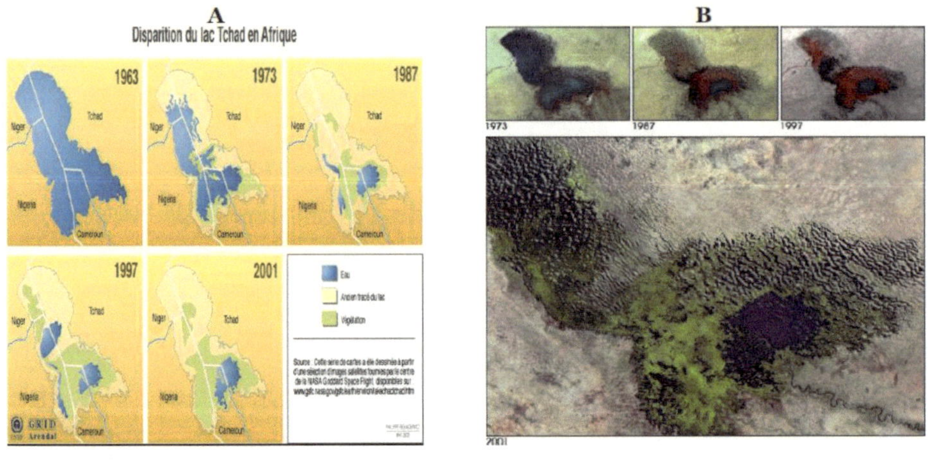

Figure 1.
Sources: NASA 2001, Internet images. Notes: (A) Changes in the surface area of Lake Chad from 1963 to 2001. (B) Satellite images of Lake Chad in 1973, 1967, 1997 and 2001.

In order to meet the Pareto criterion, these negative social consequences require compensatory transfers reflecting the loss incurred by the local population, mainly agricultural. How much should potential agricultural victims be compensated? What is the social benefit of such a project? These are two related questions that can be summarized in one main question, which is precisely the issue of this study: would the restoration of Lake Chad provide benefits that exceed the costs it would impose on farmers?

The aim of this study is precisely to answer these questions by carrying out an economic analysis of this project in order to clarify and justify the appropriateness of its implementation. Specifically, the study aims to estimate and compare the benefits and costs that the implementation of the Lake Chad Restoration Project would entail, considering only the preferences and motivations of the populations of the Cameroonian part of Lake Chad, notably the Logone and Chari departments of the Far North Region of Cameroon. This choice was mainly justified by the unavailability of time and resources to carry out a regional study including the four countries bordering the natural area which are: Cameroon, Niger, Nigeria and Chad.

The rest of this chapter is as follows: the next section sets out the analytical framework for the study. In sections 3 and 4, we carry out a monetary evaluation of the benefits and costs associated with the implementation of the project, specifying the methodological aspects specific to each part of the evaluation. Section 5 will allow us to compare these values and discusses the results obtained.

2. Analytical framework of the study

Safeguarding natural assets faces a complex dilemma to be resolved, because every choice, every option (to preserve the resource in its natural state, to let it degrade or to transform it for another use) has consequences in terms of gain or loss of value. Thus, it is only after carefully weighing all the values gained and lost under each used option that one can decide on the relevant option to be retained. For this, it is necessary to implement tools to facilitate the decision-making process. Among these tools, it is common to use cost–benefit analyses (CBA), which make it possible to shed economic light on the definition of a public objective, to make public decisions more transparent and to take into account the opinions of all the stakeholders concerned by these decisions.

In our context, a CBA could be used to justify whether or not the project to restore Lake Chad should be carried out, as it allows a monetary comparison of the costs incurred and the resulting benefits for the population. Its theoretical foundations are essentially as follows: benefits and costs are defined respectively as gains and losses in human well-being or utility [2]. Thus, a project or policy satisfies the cost–benefit criterion if its social benefits are greater than the social costs it imposes on society[1] as a whole [2–4].

However, in the case of environmental policies, the monetary evaluation of costs and benefits is difficult given the non-market nature of these goods and the services they provide. To overcome this issue, CBA assumes that individual preferences for environmental goods should be considered as the source of value. These are measured by a willingness to pay (WTP) in the case of a benefit and by a willingness to accept compensation (WTA) in the case of a cost [2]. In other words, an increase in an individual's level of well-being can be measured by the maximum amount that the individual would be willing to pay to benefit from the policy. If, on the other hand, the policy results in a reduction in well-being, it will be measured by the amount of money that the individual in question would require as compensation for accepting the policy.

Among the range of methods for valuing non-marketable environmental goods, it is common to use the contingent valuation method (CVM) to reveal these preferences [5, 6]. It is a method of valuing non-marketable goods that allows an estimate to be generated of the measures that compensate for the change in individual well-being induced by the implementation of a given public policy. Its objective is to create and simulate a hypothetical market, based on a questionnaire, in which agents are led to reveal their preferences in terms of WTP to receive a benefit and/or WTA as compensation for tolerating a given cost [7]. Using these values, it would now be possible to quantify the benefits or costs of the change one wants to make to the good concerned.

However, although CVM is the most widely used method for valuing environmental goods today [2, 5, 8], it has been strongly criticized[2], calling into question the reliability and validity of the values obtained [9]. However, instead of stopping its use, these criticisms have instead made it possible for the work to perfect it both in terms of data collection techniques and the econometric treatment of responses [10]. Although still controversial, academics and policy makers increasingly recognize this approach as a flexible and powerful method for estimating WTP [2].

However, it is important to point out three orientations that set us apart from several studies using the methods presented above. Firstly, to our knowledge, with the exception of [11], most of the work on the use of CBA to evaluate policies for safeguarding wetlands has been mainly oriented towards a comparison between the benefits and investment costs, in order to determine the appropriateness of safeguarding them without seeing their potential productive use. Then, unlike [11][3],

---

[1] In this study, as in the economic literature, the CBAs considers the costs and benefits for society as a whole, and not just for the industrialist or the community that decides whether or not to carry out a project. The aim is to assess the "social value" of the project, not its "rate of return".

[2] There are two types of criticisms: the first concerns validity, i.e. the method's capacity to estimate theoretical concepts, which leads to questions about the coherence of the results. The other criticisms relate to the reliability of the questionnaire. The latter plays a crucial role in the study.

[3] [11] Combined the use of CVM with an assessment of the benefits, at the firm level, generated by converting wetlands for industrial use.

we are interested here in the benefits of the potential agricultural use of wetlands and use CBA to measure both the benefits of restoring Lake Chad and the cost involved. Finally, we disregard the principle of discounting, which makes it possible to know whether a project is profitable by comparing different values observed at different times. Indeed, the problem that concerns us here is the difference in time between economic and ecological investments. However, economic time has nothing in common with biological or ecological time, which can extend over several hundred years. Consequently, the updating process has a clear bias against the future for long-term environmental problems, as it implies a reduced interest in the future compared to the present, and is thus a deterrent to the protection of the natural environment. Moreover, the omission of such a principle does not pose a major problem insofar as its use in the evaluation of environmental projects is still debated.

3. From the estimation of household WTP to the social benefits of the project

The monetary evaluation of the social benefits of the Lake Chad restoration project is based on two steps: the identification of the factors explaining the WTP of households wishing to participate in the Lake Chad restoration program and the calculation of the average WTP[4], which, when aggregated with the entire population concerned, enables us to obtain the said benefits.

3.1 Analysis of the factors explaining WTP

Before embarking on the analysis of these factors, we provide a brief overview of the methodology used.

3.1.1 Methodology

First of all, we present the data used. Next, the econometric model used to identify the factors relevant to explaining household WTP is presented.

3.1.1.1 Source of the data used

The data used are from a contingent survey on individual preferences for saving Lake Chad that we conducted in 2011 among households in the Cameroonian part of Lake Chad[5]. The sample is constructed through a combination of two reasoned choice survey methods, namely the quota method and the route method. The use of a probability survey requires a sampling frame (lists of households or dwellings), which unfortunately does not exist in the region, and it is impossible (limited time and means) to set it up. With this in mind, we set quotas for each itinerary and imposed on each respondent to follow a fixed itinerary (neighborhood), including stopping points for interviews. If an interviewer was absent or refused at a stopover point, the respondent moved on to the next point. This was one way to get around the problem of non-response generally observed in surveys. A total of 649

---

4 This section is a synthesis of the article published in 2018 in the Revue d'Economie Politique by [12]. For more details on the methodology and results, see this article.

5 The data collection used in this article was carried out thanks to a grant from the Research Fund for an Investment Climate and Business Environment (FR-CIEA), jointly financed by the TRUSTAFRICA Foundation and IDRC. We would like to express our sincere thanks to them.

households were asked to reveal their WTP to finance the restoration of Lake Chad through face-to-face interviews conducted in their homes. Due to the elimination of questionnaires with incomplete or missing information, 623 questionnaires are retained for the study.

The closed-ended question format with a simple dichotomous choice (yes/no) and the voluntary contribution were used in this survey as a means of formulating and revealing the WTP. The question asked, to which the respondent was asked to answer yes or no, is conveyed as follows: "*Would you agree to pay €[6]C_i per year, and for five years, to a special fund as a voluntary contribution, in the form of a donation, to ensure that Lake Chad is supplied with water?* ". The amount C_i offered to a respondent was chosen randomly from a vector of four bids (see **Table 1** below), but with a concern that each amount be distributed equitably in the total sample.

Proposed amount (€)	**1.83**	**3.66**	**9.16**	**18.32**	**Total**
Enrollment	158	155	155	155	**623**
Proportion (%)	25.36	24.88	24.88	24.88	**100**

Table 1.
Distribution of proposed amounts in the sample.

In addition to their WTP to finance the restoration of Lake Chad that could be carried out there, the respondents revealed other information about themselves: socio-economic characteristics and opinions on the safeguarding of Lake Chad. Using an econometric model, this information will allow us to test the theoretical validity of the contingent study and to predict the average WTP of our sample associated with the offer to restore Lake Chad.

Tables A1–A3 in **Annex A**, present descriptive statistics respectively on the socio-economic profile of households, the relationship between individuals and Lake Chad and the reasons for refusing to adhere to the contingent scenario.

3.1.1.2 Econometric modeling

- *Empirical specification of the econometric model*

In our situation, the respondent agrees ($Y = 1$) to participate financially in the restoration of Lake Chad if the proposed amount (C) is less than her WTP, otherwise she refuses ($Y = 0$). The discrete nature of the dependent variable leads us to model the probability that it will take the value 1 or 0. In this perspective, we choose to use the Logit model to estimate the coefficients of Eq. 1, i.e. the probability that individuals agree to pay the amount proposed in the referendum. On the basis of Pseudo R^2, the best form of regression is obtained by transforming the proposed amounts into a natural logarithm.

$$\boldsymbol{Prob}\ (\boldsymbol{Y_i = 1}) = \frac{\mathbf{1}}{\mathbf{1} + \boldsymbol{e}^{-\alpha-\beta_1 \ln(C_i) - \sum \beta_j X_j - \mu_i}} \tag{1}$$

With Y_i the dichotomous variable specifying the choice of individual i, α constant, $\beta_j (j = 1, 2, \dots, n)$ the parameters to be estimated, C_i the value of the bid proposed to individual i, β_1 coefficient of the offer (C), X_j the vector of individual explanatory variables and μ_i the error term following a logistic law.

[6] 1 euro (€) = 655 FCFA

- *Data from the econometric analysis*

The estimation procedure is performed on the data from the descriptive analysis above, from which false zeros were excluded as well as individuals who did not report their income or did not comment on the reasons for their refusal to pay the proposed amount. Thus, only information from 502 households was considered.

- *Choice of explanatory variables*

In our study, we preferred to keep in the econometric analysis the variables most correlated with WTP. There are two main reasons for this choice. On the one hand, there is no consensus on the variables influencing the decision to pay or not to pay for wetland quality improvement and their signs. On the other hand, the elimination of irrelevant variables makes it possible to obtain a more efficient econometric model [13] and to reduce the problems of multicollinearity. Thus, out of thirteen potential variables represented by the individuals' responses to the questionnaire, Chi-2 tests revealed six variables with a significant influence on WTP. The specification of these variables as well as the Chi-2 values and their associated significance thresholds are transcribed in the following **Table 2**.

Variables	Specifications	Chi-2	Prob.
	Dependent Variable		
REPONSE	= 1 if the respondent agrees to pay for the proposed offer and 0 if not.		
	Independent Variable		
Ln(OFFER)	= natural logarithm of the amount proposed in the referendum	14.776	0.002
GENDER	= 1 if the respondent is a woman and 0 if not	8.758	0.003
SIZE	= Number of individuals in the household	45.207	0.048
INCOME	= 1 if the monthly household income is greater than €152.67 and 0 otherwise	3.563	0.059
VISIT	= 1 If the respondent has already been to Lake Chad and 0 otherwise	32.849	0.000
SENSITIVITY	= 1 if the respondent is sensitive to the preservation of natural assets and 0 if not.	72.588	0.000

Table 2.
Definition of explanatory variables.

Contrary to our expectations, the variables: age, level of education, marital status, household area of residence, nationality, distance from Lake Chad and awareness of Lake Chad's shrinkage lacked a significant relationship with WTP. However, we are somewhat surprised by the non-significance of the level of education, even though several French studies [13, 14] show that it has a considerable effect on WTP. Nevertheless, the results of studies conducted in developing countries are in our view, particularly those of [15] for Tunisia or [16] for China. This non-significance of the level of education can be explained by the fact that education is not the channel of transmission that enables the local resident populations to receive the knowledge required to appreciate the value of Lake Chad.

3.1.2 Identification of the meaning of the relationship between the explanatory factors and WTP

The results of the Logit model with the significant variables are presented in **Table 3**, along with their marginal effects.

DOI: http://dx.doi.org/10.5772/intechopen.94753

Explicative Variables	Coefficient	z-stat.	P > |z|	Marginal effects
Ln(OFFER)	- 0.506*	- 3.60	0.000	- 0.074
INCOME	0.581**	2.35	0.019	0.083
GENDER	- 0.542**	- 2.10	0.035	- 0.085
SIZE	0.087*	3.35	0.001	0.012
VISIT	0.752*	3.14	0.002	0.117
SENSITIVITY	1.755*	7.01	0.000	0.316
CONSTANT	3.134*	2.58	0.010	
Pseudo R^2	0.2212	Prob> chi-2		0.0000
Log of Likelihood	- 215.91359	Number of observations	502	
Correct predictions	82.47%			

*Notes: The model has been corrected for heteroskedasticity by White's method. The signs * and ** indicate the significance of the coefficients at 1% and 5% respectively.*

Table 3.
Logistic regression results.

The results show that the model is significant (Prob> chi-2 = 0.0000) and the correct prediction rate is estimated to be 82.47%, suggesting that the model is generally well-specified. Since the estimated model has satisfactory explanatory power, let us now examine its results. With this in mind, we choose to divide the significant explanatory variables into three categories: economic, socio-demographic and behavioral.

In the economic variable category, the referendum offer and household income have a significant effect on the probability that individuals reveal a positive willingness to pay. The amount offered has a negative effect, as expected, on this probability. Thus, the probability of answering positively to the WTP evaluation question decreases as the amount offered in the referendum increases. Several studies using the CVM with dichotomous choices confirm this finding, including [15, 16], who, in their respective studies, establish a negative relationship between the proposed offer and the probability of accepting to pay the offer.

Contrary to the effect of the offer, household income positively influences the probability of agreeing to pay. As a result, households with higher incomes are also more willing to accept the amount offered than others. This result thus reflects the role that a stable and sufficient income could play in motivating individuals to participate in saving Lake Chad. This result is consistent with the finding of [15] but contradicts with [16], who instead found an insignificant relationship between income and the probability of agreeing to pay. Using a Tobit model, [14, 17] find results that are in our direction, notably a significantly positive relationship between income and respondents' WTP.

The signs of the economic variables somehow reassure the rationality of households in their response since they are in line with the expectations of economic theory [2]; indeed, the probability of accepting to pay decreases with the value of the proposed offer and increases with income. This finding allows us to theoretically validate our contingent study and to conclude that saving Lake Chad is not an inferior[7] good for the local residents.

[7] In economics, an inferior good is a good for which the income elasticity is negative. That is, a good for which demand decreases as income increases.

Concerning the socio-demographic profile of households, only two variables are significant: the gender of the respondent and the size of the household. The first has a negative effect on the probability of accepting the proposed amount. Based on this result, it can be assumed that women are less likely to pay than men. This may be explained by the fact that women do not have equivalent control over, or access to, household financial resources as men do. Thus, they are often reluctant or unable to commit the household to a substantial financial obligation. This finding is also observed by [14] who reveals that women have lower WTP than men. The second has a positive effect on the probability of accepting the bid. This result probably reflects the willingness of households to pass on this natural heritage to future generations. This result corroborates that of [15] regarding the positive significance of household size in explaining the probability of agreeing to pay for the proposed auction.

With regard to the variables related to behavior in the face of Lake Chad, the results indicate a significant positive relationship between WTP and visiting Lake Chad. Individuals who have ever visited Lake Chad are certainly aware of the threat to it, and are therefore more willing to pay to preserve it than others. This finding is consistent with the work of [11], who finds a significant and positive relationship between the WTP revealed and visiting the Seine estuary. Similarly, environmental sensitivity has a positive effect on WTP, confirming the results of [14, 17]. As a result, individuals who feel concerned about the protection or safeguarding of natural environments tend to accept the bid offered more than others. These last two results highlight not only the positive role that environmental awareness and education can play in household policies to preserve Lake Chad, but also the value of media coverage of information on the negative effects of the degradation of this natural heritage.

3.2 Average WTP and project benefits values

The main difficulty with logistic regression is that it does not allow direct measurement of average WTP. A procedure for calculating average WTP is therefore necessary.

3.2.1 Procedure for calculating average WTP

The calculation procedure we have adopted to estimate average WTP is based on Eq. 1 above and the truncation of the maximum supply from the questionnaire[8]. The expression for the corresponding average WTP[9] is:

$$\boldsymbol{Average\ WTP} = -\frac{1}{\beta_1}\left[\ln\left(1 + e^{\alpha - \beta_1 \ln(C)}\right]_0^{C^*}\right.$$

$$\Rightarrow \boldsymbol{Average\ WTP} = -\frac{1}{\beta_1}\left(\frac{1 + e^{\alpha - \beta_1 \ln(C^*)}}{1 + e^{\alpha}}\right) \qquad (2)$$

The arguments of the other variables are not explicitly apparent[10] in Eq. 2, but their influence is exerted on the values of the parameters α and β_1.

---

[8] $t_1 = 0$ *et* $t_2 = C^*$, with C^* the maximum offer of the referendum which is €18.32.

[9] See **annex C** for the procedure followed to obtain the expression of equation 2.

[10] Because they are constant and become null when the utility of the individual varies

3.2.2 Value of average WTP

Average WTP is calculated based on the econometric results of the estimation of the previous Logit model (**Table 3**). **Table 4** below gives the logarithmic value of mean WTP. Thus, the real mean WTP is obtained by applying the exponential function. The corresponding average WTP is estimated at €17 per household per year. This value is lower than those obtained in other studies carried out in developing countries and using CVM with dichotomous choices, such as those of [15, 16], who find an average WTP of €25 and $35 respectively. Also, it remains lower than French studies using payment cards: €25 for [14] or €45 for [11].

Constant (α)	**3.134**
Coefficient of the offer (β_1)	- 0.506
Ln (Average WTP)	**€0.014/year**
Average WTP	**€17/year**

Table 4.
Calculation of average WTP.

Over five years, as planned in the contingent scenario, we end up with an average WTP of €85 per household to help finance the restoration of Lake Chad. This value, aggregated across all households that value Lake Chad, allows us to obtain the benefits associated with the project.

3.2.3 Monetary evaluation of project benefits

In 2010, the population of this department was estimated at 551,718 individuals with an average of 7 people per household [18]. If we consider the national population growth rate (2.8%), we can estimate the population at 567,166 in 2011, which corresponds to an estimated 81,024 households. However, 502 households out of 623 value the preservation of Lake Chad, or 80.58% of the households in our sample.

The social benefits related to the restoration of Lake Chad are obtained by multiplying the average WTP by 80.58% of households, for an estimated monetary value of €5,549,576.832. This value represents what people in the region would be willing to give to save Lake Chad.

4. From the farmers' WTA estimate to the social costs of the project

The objective here is to estimate the farmers' willingness to receive financial compensation for the losses that would arise from the restoration of Lake Chad and to deduct the induced social cost[11]. This section is thus structured around these two main axes.

4.1 Identification and analysis of the determinants of farmers' WTA

In order to measure the farmers' willingness to receive financial compensation for the welfare losses that would result from the restoration of Lake Chad, we used the CVM.

---

[11] This section is a synthesis of the article published in 2014 in the Journal of Studies in Agriculture and Environment by [19]. For more details on the methodology and results, see this article.

4.1.1 Methodology

We successively present the data used, the econometric model used and the variables that could explain the farmers' WTA.

4.1.1.1 Data used

Data also come from the survey conducted in 2011 in the Cameroonian part of Lake Chad (presented above). This operation collected information on a sample of 98 farming households. The format of an open-ended question was used in this survey as a means of formulating and revealing the farmers' WTA[12] to bear the losses that would be related to the project's implementation.

In addition to their WTA, the interested parties revealed other information about themselves: socio-economic characteristics and opinions on the issues related to the safeguarding of Lake Chad. These indications, known as potential explanatory factors for the amounts expressed, are necessary for the econometric analysis of the latter, in order to predict an average WTA. By extrapolating the average WTA to the scale of the population concerned, this last calculation allows us to obtain the social costs resulting from the restoration of Lake Chad.

Tables A4–A6 in **Annex B** present the socio-economic profile of the farm households in our sample and their attitudes towards the project, respectively.

4.1.1.2 Econometric model

The objective of this model is to provide additional information that may help to better understand the formation of WTAs on household farmers. With this in mind, and depending on the ongoing nature of the WTA data, a simple regression model was chosen to identify the key explanatory factors for farmer WTA. The formulation of this model is as follows:

$$\mathbf{ln}\,(\boldsymbol{WTA_i}) = \boldsymbol{\beta X_i + \mu_i} \tag{3}$$

Where WTA_i is the approval to receive from individual i; β vector of parameters to be estimated, μ_i vector of error terms. Note here that preliminary regressions have shown that the semi-logarithmic model gives, in addition to reducing the potential effects of overestimating WTA values, better results than the others.

4.1.1.3 Specification of variables for the econometric analysis

The specification of the variables that may explain the value of the WTA is given in **Table 5** below.

4.1.2 Factors explaining the level of WTA

Estimation is performed on data from which false zeros have been excluded, including 7 farm households. In the end, information from 85 farm households is

[12] To obtain this value, the hypothetical scenario was conveyed as follows: "Let us assume, in a totally imaginary way, that the public authorities propose to abandon your agricultural fields located near Lake Chad in exchange for financial compensation. However, for budgetary reasons, only the inhabitants with the lowest compensation would be eligible for these subsidies. Personally, in this imaginary case, would you be willing to accept financial compensation for abandoning these fields? If so, how much would you be willing to accept to abandon these fields? ".

Variables	Specifications
STATUS	= 1 if the respondent is married and 0 if not
AGE	= 1 if the individual's age is between 30 and 60 years old and 0 if not
Ln(INCOME)	= natural logarithm of annual household farm income
SIZE	= Number of individuals in the household
DISTANCE	= 1 if the distance between the home and Lake Chad is >20 Km and 0 if not
FISHING	= 1 if the household participates in the fishing activity and 0 if not

Table 5.
Definition of the explanatory variables of the WTA.

taken into account. To this information, we applied the simplest and most immediate method of estimation, namely ordinary least squares. **Table 6** below presents the results obtained.

Explicative Variables	Coefficient	t-stat.	P > |t|
DISTANCE	- 0.176 NS	- 1.350	0.181
AGE	0.067 NS	0.900	0.369
STATUS	0.209*	1.860	0.067
SIZE	0.013*	1.730	0.088
FISHING	- 0.027 NS	- 0.280	0.781
LN(INCOME)	0.684***	10.600	0.000
CONSTANT	3.677***	4.210	0.000
R^2	0.7314		
F-Stat.	29.84		
Prob (F-Stat.)	0.0000		
Number of observations		85	

*Notes: The model has been corrected for heteroskedasticity by White's method. The ***, **, and * indicate the significance of the variables at thresholds of 1, 5, and 10 per cent, respectively. NS: Not significant.*

Table 6.
Results of the estimation of Eq. 3.

The results presented in **Table 6** above call for the following comments.

First of all, the WTA revealed by individuals is not a randomly announced value, but depends mainly on the level of farm income, probably reflecting the absence of strategic bias such as the desire to block the project or the announcement of compensation amounts that are totally disproportionate to the actual costs incurred by the project. The WTAs obtained should therefore be considered normal, as they are formulated in relation to the agricultural losses incurred. This result thus contradicts those often put forward in the literature, notably the abnormally high or infinite nature of the WTA values [20] or the popular conclusion that budgetary constraint does not play a role in the revelation of the WTA, but remains consistent with those of [21–23] who find a statistically significant and positive relationship between the levels of compensation requested by farmers and farm income.

Second, the positive sign of the coefficient indicates that households with high farm incomes tend to demand higher levels of compensation than others. This result is somewhat reassuring to the rationality of the households in their response, and is indicative of the close relationship between farm income and the level of WTA.

All factors being equals, the calculation of the income elasticity of the WTA shows that a 10% increase in income would result in a 6.84% increase in WTA, confirming the predominant role of agricultural income in the formulation of the WTA.

Finally, the results highlight the positive roles of marital status and household size on the WTA level. Couple households would thus require higher compensation than other households. Similarly, households with many members tend to express higher WTAs than others. These results can be partly explained by the strong tendency of these households to engage in agricultural activities [24], and therefore would be more vulnerable to the induced effects of the project. They also point to a greater loss of income for these farming households than would be induced by the project, and therefore suggest that any compensation should be modulated according to these criteria.

4.2 Econometric estimation of the average WTA and deduction of social costs

4.2.1 Econometric estimation of the average WTA

Econometric modeling provides, based on the estimated coefficients, the predicted WTA values for all individuals in the sample, even for those not included in the regression. The average WTA resulting from this modeling is thus estimated at €1,405.515 within a 95% confidence interval of [1,246.803; 1,564.226].

4.2.2 Estimating the social costs of restoring Lake Chad

The social costs associated with the Lake Chad restoration project are obtained by multiplying the average WTA by the number of potential victims. However, during data collection, it was not possible to determine the exact number of farming households that could be affected by the project. Therefore, this number was estimated based on the total agricultural population of the Cameroonian districts sharing Lake Chad, i.e., 27,008[13] agricultural households in 2010.

According to the data collected, the social costs are estimated at €37,960,149.12 with a 95% confidence interval of [33,673,656.87; 42,246,614.57]. This value represents the monetary assessment of the damage that the Lake Chad restoration project would impose on the riparian agricultural population.

5. Comparison of values and discussion

The average WTP is estimated (in the medium term) to be €85 per household, or a total benefit of €5,549,576.832 from the Lake Chad restoration project. On the other hand, the average compensation to be paid to potential victims is in the order of €1,405.515, so the total cost of the project would be around €37,960,149.12[14]. It thus appears that the costs of restoring Lake Chad clearly outweigh the benefits. This result is in line with the conclusions of [11, 25] on the trade-off between productive/non-productive uses of wetlands, but contradicts the work of [22], which highlights the superiority of the benefits (12.6 million euros) over the costs (2.3 million euros) for the preservation of 2000 hectares (ha) of natural habitats along the Garonne River. This difference in results can nevertheless be explained by the observed differences in the sizes of the beneficiary populations (250,000

[13] Data collected at the MINADER departmental delegation.

[14] It should be noted that this amount should be used with caution, as it takes into account all farm households in the districts concerned, and not the potential victims of the project.

households) and their standards of living. For the average WTP is around €50.4/year and the average WTA around €1150/ha/year.

We thus find ourselves in a paradoxical situation in which, in the medium term (5 years), the agricultural conversion benefits of Lake Chad are far greater than what the people of the region will be willing to give to save it. This result would therefore justify the trade-off to be made in favor of a productive use of Lake Chad's wetlands. However, this difference must be tempered given:

- The omission of negative externalities (water pollution, deforestation, etc.) in the calculation of the benefits induced[15] by agricultural activities;
- The overestimation of costs due to the impossibility of counting the potential victims of the project;
- The low representativeness of the sample[16] in relation to the total population living along the shores of Lake Chad;
- The time horizon: economic or ecological. For an economic horizon (5 to 10 years), it is not profitable to carry out the project. On the other hand, if the ecological horizon is chosen (from 35 years), it becomes profitable to carry out the project, because the benefit/cost ratio would then become greater than 1, since the benefits of the project would then be around €38,543,518.56.

The amounts announced are therefore indicative for all policy choices, and must therefore be taken with caution. That said, the results obtained can nevertheless serve as a basis for deliberation among the parties concerned, without being a sufficient or necessary condition for the decision to implement a policy to restore Lake Chad. Other reasons may justify this policy.

6. Conclusion

At the end of our study, it is very clear that the protection of Lake Chad is of crucial necessity, given the economic and environmental importance of this natural area. In economic terms, Lake Chad is much more than a source of recreation for its residents; it is the very foundation of their food security. It guarantees the economic and social dynamics of the region and constitutes, in a way, a pool of jobs linked to the economic activities that are found around it. From an environmental standpoint, the survey also reveals that Lake Chad is perceived by its neighbors and farmers as a true natural heritage and, as such, should be saved from its likely fate. To this effect:

- 73.51% of the riparian populations are in favor of the Lake Chad restoration project, and 63.40% would not hesitate to allocate a certain portion of their income to participate financially in the realization of such a project.

[15] Taking externalities into account would significantly reduce the value of productive benefits. However, to our knowledge, no research has yet focused on the monetary quantification of these externalities. Yet, such an evaluation would make it possible to determine the net benefits of Lake Chad's agricultural use.

[16] The absence of a part of the population concerned deprives us of useful information on a possible contribution of this layer to the evaluation of the project. However, the transferability of values could lift this limit.

- Despite their dependence on the productive services it provides, 70% of riparian farmers are in favor of the project. This percentage rises to about 87% when financial compensation is offered in return.

Its statistics are thus proof that, from the population's point of view, there is a real benefit in preserving Lake Chad, even if the estimate of this benefit appeared low, or at least lower than that generated by the agricultural use of Lake Chad's shores. However, this difference does not mean that the economy should be favored over the environment, but it does make it possible to highlight the possible perverse effects that could result from carrying out the Lake Chad restoration project without taking into account its potentially negative impact on agricultural activities, and thus on the food security of the riparian populations. In view of the above, we propose a new approach to save Lake Chad from its probable disappearance, one that takes into account both economic and environmental objectives. Thus, rather than opposing the quality of the environment to the pursuit of economic development, we believe that these two objectives are in close interaction and that a policy to save Lake Chad should ideally take into account all of them rather than focusing on one or the other. The economic literature speaks of sustainable development.

Furthermore, we believe that restoring Lake Chad without addressing the causes that led to its dwindling could be an unsustainable solution because the same causes will produce the same effects. Indeed, even if this natural asset were to be restored, there is no guarantee that it will not dwindle again and endanger the river that provides it.

Thus, whatever the choice of decision-makers, it is essential to implement policies for the protection of this natural heritage. To this end, it is crucial to move from logic of supply management of the natural resources of the lake basin to logic of demand management of the said resources. This can be achieved by putting in place incentive mechanisms for the sustainable management of the wetlands of Lake Chad and the biodiversity associated with them.

Annex

Annex A. Some descriptive statistics of the WTP Revelation Scenario

Socio-Economic Profile		Enrolled	Percentage (%)
SEX	Male	436	69.98
	Female	187	30.02
AGE	Average of 30 years	187	30.02
	From 30 to 44 years	185	29.70
	From 45 to 60 years	208	33.39
	Above 60 years	43	6.90
MARRITAL STATUS	Married	321	51.52
	Divorced / Widow/Widower	118	18.94
	Single	184	29.53
LEVEL OF EDUCATION	None	151	24.24
	Primary	77	12.36
	Secondary 1st cycle	74	11.88

Socio-Economic Profile		Enrolled	Percentage (%)
	Secondary 2nd cycle	177	28.41
	Tertiary	102	16.37
	Not stated	42	6.74
AVERAGE MONTHLY INCOME	Less than €30.53	75	12.04
	From €30.54 to €76.34	147	23.60
	From €76.35 to €152.67	116	18.62
	From €152.68 to €305.34	97	15.57
	From €305.35 to €763.36	99	15.89
	Above €763.36	59	9.47
	Is not pronounced	30	4.82

Table A1.
Sample distribution by socio-economic profile.

	Number of observations	Percentage (%)
Sensitivity to the protection of natural environments	414	66.45
Visits to Lake Chad	382	61.32
Awareness of Lake Chad dwindling	499	80.10
Favorable to the Lake Chad Restoration Project	458	73.51
Agrees to pay the proposed amount	395	63.40

Table A2.
Relationship and preferences of individuals towards Lake Chad.

Reasons	Type of zero	Enrolled	Proportion
It's not up to you to pay	False zero	51	22.37%
You do not consider this action necessary	True zero	66	28.95%
Your financial means do not allow you to do so	True zero	51	22.37%
You do not have enough information to decide	False zero	25	10.96%
You are afraid to pay for others	False zero	13	5.70%
Other reasons	False zero	14	6.14%
No opinion	False zero	8	3.51%
Total		228	100%

Table A3.
Reasons for refusing to pay the proposed amount.

Annex B. Descriptive statistics of the Farmers' WTA Revelation Scenario

Socio-Economic Profile	Enrolled	Proportion
Average size of household	9.74	

Socio-Economic Profile		Enrolled	Proportion
Civil status	Married	77	78.57%
	Divorced	13	13.27%
	Single	8	8.16%
Sex	Male	86	87.76%
	Female	12	12.24%
Age	Age < 30	19	19.39%
	30 ≤ Age < 45	33	33.67%
	45 ≤ Age ≤ 60	25	25.51%
	Age > 60 years	21	21.43%
Level of education	None	53	54.08%
	Primary	32	32.65%
	Secondary 1er cycle	9	9.18%
	Secondary 2nd cycle	4	4.08%

Table A4. Sample distribution by socio-economic profile.

Surface area in hectares (ha)	Enrolled	Proportion	Average farm income (€)
Surface area < 1	4	4.08%	885.50
1 ≤ *Surface area* < 2	15	15.31%	1,548.09
2 ≤ *Surface area* < 5	18	18.37%	2,256.49
5 ≤ *Surface area* < 10	32	32.65%	2,297.71
Surface area ≥ 10	29	29.59%	3,734.35
TOTAL	**98**	100%	2,144.43

Table A5. Attribution of average farm income by surface area.

		Enrolled	Proportion
Opinion of the project	Very important	15	15.31%
	Important	54	55.10%
	Less important	26	26.53%
	Not important at all	03	3.06%
Favorable for compensation	Yes	85	86.73%
	No	13	13.27%
Reasons for Refusal	**Type of zero**		
I do not feel concerned	True zero	0 6	46.15%
I want to continue my activities	False zero	05	38.46%
This action is not necessary	False zero	02	15.38%

Table A6. Attribution according to individuals' attitudes towards the fictitious scenario.

Annex C. Procedure for obtaining Equation 2

Assume that the individual's utility depends on improvement in the quality of a good's environment q, its income R and other observable socio-demographic variables m. Based on these hypotheses, we retain the linear utility function as follows:

$$v(q, R, m) = \alpha_q + \beta R + \gamma m \qquad q = 0, 1 \tag{A1}$$

In a closed question, the respondent agrees to pay an amount of money C in order to benefit from the restoration of Lake Chad, if

$$v(1, R - C, m) \geq v(0, R, m) \tag{A2}$$

The induced variation in utility (Δv) is therefore a function of cost C:

$$\begin{aligned} \Delta v = \alpha_1 + \beta(R - C) + \gamma m - \alpha_0 - \beta R - \gamma m = (\alpha_1 - \alpha_0) - \beta C \\ \alpha_1 - \alpha_0 = \alpha \quad \Longrightarrow \quad \Delta v = \alpha - \beta C \end{aligned} \tag{A3}$$

Let F be the distribution function of a logistic law, then the probability (P_0) that individuals are willing to pay the amount C can be expressed by the following equation:

$$P_0 = F_\varepsilon(\alpha - \beta C) = \frac{1}{1 + e^{-\alpha + \beta C}} = \frac{e^{\alpha - \beta C}}{1 + e^{\alpha - \beta C}} \tag{A4}$$

From this equation (4), the average WTP is defined by:

$$\begin{aligned} \text{Average WTP} = E(WTP) = \int_{t_1}^{t_2} F_\varepsilon(\alpha - \beta C) dC \\ \Longrightarrow E(WTP) = \int_{t_1}^{t_2} \frac{e^{\alpha - \beta C}}{1 + e^{\alpha - \beta C}} dC \end{aligned} \tag{A5}$$

In order to calculate this integral, it is necessary to look for a primitive of the function which is inside the integral.

First of all, let's assume:

$$u(C) = 1 + e^{\alpha - \beta C} \quad \Rightarrow \quad u'(C) = -\beta e^{\alpha - \beta C} \tag{A6}$$

Next, let's look for:

$$E(WTP) = \int_{t_1}^{t_2} \frac{u'(C)}{u(C)} dC \;\Leftrightarrow\; E(WTP) = [ln\,(u(C))]_{t_1}^{t_2} \tag{A7}$$

In this perspective, we will multiply the numerator and denominator of the function within the integral by $-\beta$. We have:

$$\begin{aligned} E(WTP) = \int_{t_1}^{t_2} \frac{-\beta}{-\beta} * \frac{e^{\alpha - \beta C}}{1 + e^{\alpha - \beta C}} dC = \int_{t_1}^{t_2} -\frac{1}{\beta} * \frac{-\beta e^{\alpha - \beta C}}{1 + e^{\alpha - \beta C}} dC \\ \Longrightarrow E(WTP) = -\frac{1}{\beta} \int_{t_1}^{t_2} \frac{-\beta e^{\alpha - \beta C}}{1 + e^{\alpha - \beta C}} dC \\ \Leftrightarrow E(WTP) = -\frac{1}{\beta} \int_{t_1}^{t_2} \frac{u'(C)}{u(C)} dC = -\frac{1}{\beta} [ln\,(u(C))]_{t_1}^{t_2} \\ \Longrightarrow E(WTP) = -\frac{1}{\beta} \left[ln\,\left(1 + e^{\alpha - \beta C}\right)\right]_{t_1}^{t_2} \end{aligned} \tag{A8}$$

Where:

$$E(WTP) = \int_{t_1}^{t_2} \frac{e^{\alpha-\beta C}}{1+e^{\alpha-\beta C}} dC = -\frac{1}{\beta}\left[ln\left(1+e^{\alpha-\beta C}\right)\right]_{t_1}^{t_2}$$

Author details

Patrick Arnold Ombiono Kitoto
The University of Maroua, Maroua, Cameroon

*Address all correspondence to: ombionopatrick@yahoo.fr

References

[1] Ombiono Kitoto P. A. Réchauffement climatique et migration vers les rives du lac Tchad. *Migrations société*. 2016 ; 28 (163) : 149–165.

[2] Pearce D, Atkinson G, Mourato S. *Analyse Coûts-Bénéfices et Environnement : Développements Récents*. OCDE ; 2006.

[3] Bouyssou D, Marchant T, Pirlot M, Perny P, Tsoukiàs A, Vincke Ph. *Evaluation and decision models : a Critical Perspective*. Kluwer Academic Publishers; 2000.

[4] Meunier V. Analyse coût-bénéfices : guide méthodologique. *Cahiers de la Sécurité Industrielle*. 2009 ; 06. Available at http://www.icsi-eu.org/francais/dev_cs/cahiers/.

[5] Arrow K, Solow R, Learner E, Portney P, Randner R, Schuman H. Report of the NOAA Panel on Contingent Valuation. *U. S. Federal Register*. 1993; 10: 4601–4614.

[6] Wilson JJ, Lantz VA, MacLean DA. A benefit-cost analysis of establishing protected natural areas in New Brunswick, Canada. *Forest Policy and Economics*. 2010; 12: 94–103.

[7] Faucheux S, Noël JF. *Economie des Ressources Naturelles et de l'Environnement*. Armand Colin; 1995.

[8] Mitchell RC, Carson RT. *Using Surveys to Value Public Goods: the Contingent Valuation Method*. Resources for the future; 1989.

[9] Hausman JA. *Contingent Valuation: a critical assessment*. Hausman JA; 1993.

[10] Willinger M. La méthode d'évaluation contingente : de l'observation à la construction des valeurs de préservation. *Natures-Sciences-Sociétés*. 1996 ; 4 : 6–22.

[11] Beaumais O, Laroutis D, Chakir R. Conservation versus Conversion des Zones Humides : Une Analyse Comparative Appliquée à l'Estuaire de la Seine. *Revue d'Economie Régionale et Urbaine*. 2008 ; 4. 565–590.

[12] Ombiono Kitoto P A. Evaluation du consentement à payer des riverains pour la restauration du lac Tchad. *Revue d'économie politique*. 2018 ; 128(2) : 1177–1200.

[13] Raboteur J, Divialle F. Mangrove et développement touristique à Port-Louis : évaluation économique d'un espace naturel humide de Guadeloupe. In: Roselé Chim P, Raboteur J, directeurs. *Les défis du développement du tourisme et l'intelligence économique des territoires : villes sur mer et en terre : actes du colloque des 3 et 4 décembre 2009*. Publibook ; 2014. p. 85–114.

[14] Bonnieux F, Le Goffe P, Vermersch D. La méthode d'évaluation contingente : application à la qualité des eaux littorales. *Economie et Prévision*. 1995 ; 117–118 : 89–106.

[15] Belloumi M, Matoussi MS. Evaluation de la valeur de préservation de la qualité de la nappe d'Oued Kheirate. *New Medit*. 2002; 4:39–45.

[16] Shang Z, Che Y, Yang K, Jiang Y. Assessing local communities' willingness to pay for river network protection: a contingent valuation study of Shanghai, China. *International Journal of Environmental Research and Public Health*. 2012 ; 9(11):3866–3882.

[17] Voltaire L, Nassiri A, Bailly D, Boncoeur J. Effet d'une taxe et d'un droit d'entrée sur les consentements à payer des touristes pour de nouvelles réserves naturelles dans le Golfe de Morbihan. *Revue d'études en agriculture et environnement*. 2011;92 (2): 183–209.

[18] MINADER. *Campagne Agricole 2010*. Rapport annuel d'activité de la délégation départementale du Logone et Chari, Ministère de l'Agriculture et du Développement Rural, Cameroun. 2010.

[19] Ombiono Kitoto PA. Evaluation du consentement à recevoir des agriculteurs pour la restauration du lac Tchad. *Revue d'Etudes en Agriculture et Environnement*. 2014;95(4):493–512.

[20] Angel M. *La nature a-t-elle un prix ?*. Les Presses de l'Ecole des Mines ; 1998.

[21] Amigues JP, Desaigues B. L'évaluation d'une politique de protection de la biodiversité des forêts riveraines de la Garonne. Chapitre 3, in Point P. (dir). La valeur économique des hydrosystèmes. Méthodes et modèles d'évaluation des services délivrés. *Economica* ; 1999.

[22] Amigues JP, Boulatoff C, Desaigues B, Gauthier C, Keith JE. The Benefits and Costs of Riparian Analysis Habitat Preservation: A Willingness to Accept/Willingness to Pay Contingent Valuation Approach. *Ecological Economics*. 2002;43:17–31.

[23] Buckley C, Hynes S, Mechan S. Supply of an ecosystem service: Farmers' willingness to adopt riparian buffer zones in agricultural catchments. *Environmental Science and Policy*. 2012; 24:101–109.

[24] Ombiono Kitoto PA. *Déterminants de la polarisation des activités agricoles dans le bassin du lac Tchad*. In Tsafack NR, Khan SA, éditeurs. Environnement des Affaires et Systèmes Productifs en Afrique Centrale. Africaine d'Edition ; 2013.

[25] Seung CK, Harris TH, Englin JE. Impacts of Water Reallocation: A Combined Computable General Equilibrium and Recreation Demand Model Approach. *The Annals of Regional Science*. 2000;34:473–487.